Springer Theses

Recognizing Outstanding Ph.D. Research

Aims and Scope

The series "Springer Theses" brings together a selection of the very best Ph.D. theses from around the world and across the physical sciences. Nominated and endorsed by two recognized specialists, each published volume has been selected for its scientific excellence and the high impact of its contents for the pertinent field of research. For greater accessibility to non-specialists, the published versions include an extended introduction, as well as a foreword by the student's supervisor explaining the special relevance of the work for the field. As a whole, the series will provide a valuable resource both for newcomers to the research fields described, and for other scientists seeking detailed background information on special questions. Finally, it provides an accredited documentation of the valuable contributions made by today's younger generation of scientists.

Theses are accepted into the series by invited nomination only and must fulfill all of the following criteria

- They must be written in good English.
- The topic should fall within the confines of Chemistry, Physics, Earth Sciences, Engineering and related interdisciplinary fields such as Materials, Nanoscience, Chemical Engineering, Complex Systems and Biophysics.
- The work reported in the thesis must represent a significant scientific advance.
- If the thesis includes previously published material, permission to reproduce this must be gained from the respective copyright holder.
- They must have been examined and passed during the 12 months prior to nomination.
- Each thesis should include a foreword by the supervisor outlining the significance of its content.
- The theses should have a clearly defined structure including an introduction accessible to scientists not expert in that particular field.

More information about this series at http://www.springer.com/series/8790

Tuncay Ozel

Coaxial Lithography

Doctoral Thesis accepted by
Northwestern University, Evanston, USA

Springer

Author
Dr. Tuncay Ozel
Department of Chemistry and Chemical Biology
Harvard University
Cambridge, MA
USA

Supervisor
Prof. Chad A. Mirkin
Department of Chemistry
Northwestern University
Evanston, IL
USA

ISSN 2190-5053 ISSN 2190-5061 (electronic)
Springer Theses
ISBN 978-3-319-45413-9 ISBN 978-3-319-45414-6 (eBook)
DOI 10.1007/978-3-319-45414-6

Library of Congress Control Number: 2016950377

Printed on acid-free paper

This Springer imprint is published by Springer Nature
The registered company is Springer International Publishing AG
The registered company address is: Gewerbestrasse 11, 6330 Cham, Switzerland

This dissertation is dedicated to my families

Supervisor's Foreword

Nanoscale structures offer unlimited opportunities in advancing our knowledge of science and technology because of their unique physical and chemical properties. Such properties are tailorable by varying nanoscale architecture, and recent advances in lithographic and synthetic techniques have enabled the realization of a variety of fundamentally interesting, complex nanoscale systems. Particles prepared using high-resolution methods such as photolithography, electron-beam lithography, dip-pen lithography, nanoimprint lithography, and on-wire lithography have led to the discovery of important structure–function relationships in the context of optics, electronics, and catalysis. Novel materials prepared using such precise structural control have resulted in significant progress in solar to chemical energy conversion and storage, biological agent detection, drug delivery, and multifunctional electronic devices.

Although there are now reliable methods for synthesizing metallic structures with well-defined, and in certain cases exotic, geometries, most lithographic systems suffer from poor control over the radial dimension. Such architectural control is essential to synthesize plasmonically and catalytically active metallic nanostructures of appropriate geometric dimensions in and around coaxial semiconductor nanowires. Such structural motifs could serve as foundational components for next-generation photovoltaic and photocatalytic systems. This thesis addresses this challenge by describing the development of a new technique, coaxial lithography (COAL), which bridges templated electrochemical synthesis and lithography. This technique uniquely allows controlled generation of coaxial nanowires with compositional freedom and sub-10 nm resolution. Synthetic control over the radial dimension, combined with the possibility of selectively deleting features within the design paradigm, significantly expands the range of nanoparticle architectures that can be synthesized using COAL. In particular, the integration of metal nanorings around and within semiconductor nanowires with unprecedented control over their locations and dimensions is demonstrated. The capacity to integrate these two types of materials into one construct is heavily sought after due to the extraordinary ability of metal nanostructures to enhance light absorption within semiconductors.

As a proof of concept, enhancement of photocurrent from coaxial organic/inorganic core/shell semiconductor nanowires was studied by controllably embedding plasmonic metal nanorings within the nanowires without blocking the charge carrier flow.

In sum, this work represents a significant advance in nanomaterials synthesis by design because it is applicable to a wide variety of materials (metals, organic semiconductors, metal oxides, and metal chalcogenides). As such, this thesis will have a substantial impact in the fields of chemistry, physics, materials science, nanotechnology, and many other scientific disciplines, as it allows for the synthesis of materials with architectures previously unachievable through other means.

Evanston, USA
April 2016

Prof. Chad A. Mirkin

Abstract

The optical and electrical properties of heterogeneous nanowires are profoundly related to their composition and nanoscale architecture. However, the intrinsic constraints of conventional synthetic and lithographic techniques have limited the types of multi-compositional nanowires that can be realized and studied in the laboratory. This thesis focuses on bridging templated electrochemical synthesis and lithography for expanding current synthetic capabilities with respect to materials generality and the ability to tailor two-dimensional growth in the formation of core-shell structures for the rational design and preparation of nanowires with very complex architectures that cannot be made by any other techniques.

Chapter 1 introduces plasmonics, templated electrochemical synthesis, and on-wire lithography concepts and their significances within chemistry and materials science.

Chapter 2 details a powerful technique for the deposition of metals and semiconductors with nanometer resolution in segment and gap lengths using on-wire lithography, which serves as a new platform to explore plasmon–exciton interactions in the form of long-range optical nanoscale rulers.

Chapter 3 highlights an approach for the electrochemical synthesis of solution dispersible core-shell polymeric and inorganic semiconductor nanowires with metallic leads. A photodetector based on a single core-shell semiconductor nanowire is presented to demonstrate the functionality of the nanowires produced using this approach.

Chapter 4 describes a new materials general technique, termed coaxial lithography (COAL), bridging templated electrochemical synthesis and lithography for generating coaxial nanowires in a parallel fashion with sub-10 nanometer resolution in both axial and radial dimensions. Combinations of coaxial nanowires composed of metals, metal oxides, metal chalcogenides, conjugated polymers, and a core/shell semiconductor nanowire with an embedded plasmonic nanoring are presented to demonstrate the possibilities afforded by COAL.

Chapter 5 addresses the use of COAL for the synthesis of solution dispersible metal nanorings and nanotubes with exceptional architectural tailorability of inner diameter, outer diameter, and length leading to precise spectral control over the resulting plasmonic fields ranging from visible to the near-IR.

Chapter 6 is an outlook on templated electrochemical synthesis using coaxial lithography and highlights a few promising applications from nanoparticle assembly to light–matter interactions.

Tuncay Ozel
Thesis Advisor: Prof. Chad A. Mirkin

Parts of this thesis have been published in the following journal articles:

Nano Letters 15 (8), 5273–5278 (2015).
Nature Nanotechnology 10 (4), 319–324 (2015).
Advanced Materials 25 (32), 4515–4520 (2013).
Nano Letters 13 (5), 2270–2275 (2013).
Nano Letters 12 (12), 6218 (2012).

Acknowledgments

I am grateful to Prof. Chad A. Mirkin for giving me the opportunity to work in his rewarding lab, being an inspirational role model scientist for me, and his endless support during my Ph.D. studies. I feel very lucky to have learned the Mirkin group culture, which has helped me in making substantial progress towards my goal of being a scientist.

I would like to thank my thesis committee, Profs. George C. Schatz, Mark C. Hersam, and Vinayak P. Dravid for their guidance and helpful discussions. In particular, I would like to thank Prof. Schatz for his mentorship and support in our collaborative research works. His advice and suggestions helped me enormously since my first interaction with him. I would like to thank Prof. Lincoln J. Lauhon for his support during my Ph.D. studies at Northwestern, and also my former adviser Prof. Hilmi Volkan Demir from Bilkent University for first introducing me to scientific research and for his endless support throughout my academic career.

I had a wonderful experience in the Mirkin group with the opportunity to interact with outstanding researchers in a very collaborative and friendly setting. First of all, I would like to thank many times my close collaborator and lifelong friend Dr. Gilles R. Bourret. Most of the scientific breakthroughs that are presented in this thesis wouldn't be possible without his help, wisdom, and enthusiasm. I want to thank Mike Ashley and Taegon Oh for their contributions to my research during my last year at Northwestern University. I believe that these two bright young graduate students will be extremely successful in their Ph.D. studies and post-Northwestern. I would like to thank all former and current members of the Mirkin Group. I had a chance to meet a lot of incredible people, build amazing friendships, and have fruitful collaborations during last 4 years. In particular, I want to thank my wonderful friends Dr. Bryan Mangelson, Jessie Ku, Matthew O'Brien, Lam-Kiu Fong, Dr. Abrin Schmucker, Dr. Matthew Jones, Dr. Keith Brown, Dr. Kyle Osberg, Michael Ross, Ryan Thaner, Youngeun Kim, Eileen Soyoung Seo, Daniel Park, Dr. Evelyn Auyeung, Dr. Jennifer McGinnis, Dr. Sarah Hurst, Alejo Lifschitz, Mike McGuirk, and Resham Banga for their great support and contributions.

I also would like to thank Taner Aytun, Deep Jariwala, Besim Avci, Kezban-Gurkan Sokat, Faifan Tantakitti, Bernard Beckerman, Yigit Demir, Emre-Yeliz Karaman, and Servet Seckin Senlik who have made the entire graduate school experience more enjoyable.

I want to thank my parents, Ali and Yildiz Ozel, for their endless love and support, which has helped me tremendously in achieving my goals since my childhood. I want to thank my family members Taner-Zulal Ozel and Ayhan-Simge-Ulas Huyal for their support and trust in me. Lastly, I would like to thank my lovely wife Ozge Ozel who always wanted the best for me and encouraged me to do so. Thank you for your patience and support during my graduate studies in which we have experienced living a thousand miles away from each other. Finally it is over, and I am looking forward to enjoying our future together.

Evanston, USA
January 2015

Tuncay Ozel

Contents

List of Figures

Chapter 1
Introduction to Plasmonics, Templated Electrochemical Synthesis, and On-Wire Lithography

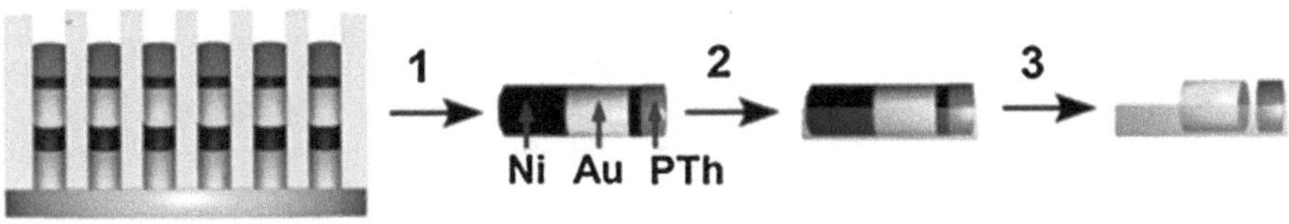

1.1 Motivation and Introduction

High-resolution synthetic and lithographic tools, enabling excellent control of material composition and nanoscale architecture, are required to manipulate and tailor the properties of metals and semiconductors [1–3]. Research in physics, chemistry, and materials science are highly dependent on the development of these technologies [1–6]. For instance, recent improvements in synthetic tools have led to remarkable progress in the understanding of how metal nanostructures interact with light in a number of metal-metal [1, 6–9] metal-semiconductor [4, 10–14] and metal-insulator [1, 15–17] platforms, and have also led to their use in a range of applications from surface-enhanced spectroscopies [18, 19] to plasmonic metamaterials [1, 20]. In particular, metallic nanostructures placed in proximity to semiconductors were discovered to act as antennae that transfer energy through

Portions of this chapter have been published in ***Nano Letters*** 12 (12), 6218 (2012).

Co-authors of this work: Matthew Rycenga, Mark R. Langille, Michelle L. Personick, Chad A. Mirkin.

T. Ozel, *Coaxial Lithography*, Springer Theses,
DOI 10.1007/978-3-319-45414-6_1

surface plasmon resonances to enhance light absorption by the semiconductor [2, 4, 10, 11, 21]. Such structures have allowed researchers to explore novel pathways for energy harvesting and molecular sensing, and they have been proposed as promising entities for improving the conversion of incident solar radiation into electrical energy [4, 14, 22]. Similarly, the use of coaxial nanowires composed of optically active p- and n-type semiconductors were shown to improve the conversion of photons into electrical and chemical energy due to radial charge separation [22–25]. Though there are a number of methods used to generate coaxial nanowires (such as atomic-layer deposition, sputtering, chemical vapor deposition, and layer-by-layer assembly), these methods lack the ability to engineer nanostructures along multiple directions (i.e. precise control of the composition, length, diameter, and position of multiple segments). This intrinsic constraint of conventional synthetic and lithographic techniques has limited the types of multi-compositional nanowires that can be realized and studied in the laboratory. For example, metal nanorings have emerged as important nanoscale materials due to their exceptional optical and magnetic properties, however, they cannot be integrated to specific sites along coaxial nanowires using conventional fabrication tools. Such structures could, for example, enhance light emission and absorption within a semiconductor nanowire by generating an intense electric field localized inside the ring. This thesis focuses on the synthesis of a wide variety of materials (metals, organic semiconductors, metal oxides and metal chalcogenides) with architectures unachievable via other means to advance nanomaterials synthesis by design. As such, this thesis is intended to contribute to the fields of chemistry, physics, materials science, and nanotechnology.

1.2 Plasmonic Properties of Metal Nanoparticles

The synthesis of metal nanoparticles has remained a major field of research in the last few decades since such particles show extraordinary optical properties compared to their bulk counterparts [1, 6, 17, 26]. For example, if the size of a metal nanoparticle is much smaller than the wavelength of an incident light source, for some noble metals, light can penetrate into the nanoparticle and polarize charges on the metal surface (Fig. 1.1a) [3, 27]. This polarization drives delocalized electrons out of the metal surface, but the Coulombic attraction between the positively charged nucleus and the negatively charged conduction band electrons (described as a free electron cloud) creates a restoring force which pulls the electron cloud back to its initial position. Since the direction of the polarization oscillates (as a sinusoidal electromagnetic wave) in a light source, the consequent restoring force moves the electrons back and forth to balance this oscillation in the polarization of the light source. Finally, collective oscillation of the conduction band electrons and the excitation source (causing the greatest amount of charge displacement when they are oscillating in resonance) generates localized and intensified electric fields (100 to 10,000 times the incident field) around the metal nanoparticle (known as localized plasmons) and the frequency of this oscillation is called the localized surface plasmon resonance (SPR) frequency [3, 27, 28]. For example, a 10 nm Au nanosphere can support localized plasmons in which intense and localized electric fields can be visualized in the calculated electric field intensity map of this nanoparticle (Fig. 1.1b). Electric field intensity is the greatest on the surface of nanoparticles (oriented in the same direction as the polarization of the light source) and decays away into the dielectric environment. This electric field intensity map was simulated with an excitation source propagating along the x-axis and polarized along the y-axis (using a commercially available finite-difference time-domain Maxwell equations solving software package by Lumerical Solutions Inc., Vancouver, Canada).

SPR frequency of a metal nanoparticle strongly depends on the size, shape, composition, and the surrounding dielectric medium [27]. For example, Au and Ag nanoparticles show localized plasmon resonance wavelengths in the visible, whereas Al has a higher oscillation frequency corresponding to a resonance wavelength in the UV [3, 27, 29]. Another important factor in the plasmon resonance frequency is the size of the nanoparticle such that the SPR frequency of a nanoparticle gets smaller for larger nanoparticles. This relation can be expressed as:

$$\omega = \left(\frac{e^2}{\varepsilon_0 m_e 4\pi r_s^3}\right)^{1/2} \tag{1.1}$$

where ε_0 is the vacuum permittivity, m_e is the effective mass, and r_s is radius of the sphere [30]. Experimental efforts with Au and Ag nanoparticles confirm a redshift in the plasmon resonance wavelength with increasing particle size [31–33]. Another approach to modify the surface plasmon resonance frequency of a nanoparticle is to change its surrounding dielectric medium [34]. For example, the surface plasmon resonance wavelength of gold nanoparticles redshifts when they are covered with a silica shell, since silica has a higher refractive index compared to water or air (refractive index of silica: 1.45, water: 1.33, air: 1.00) [34]. Lastly, there is a strong

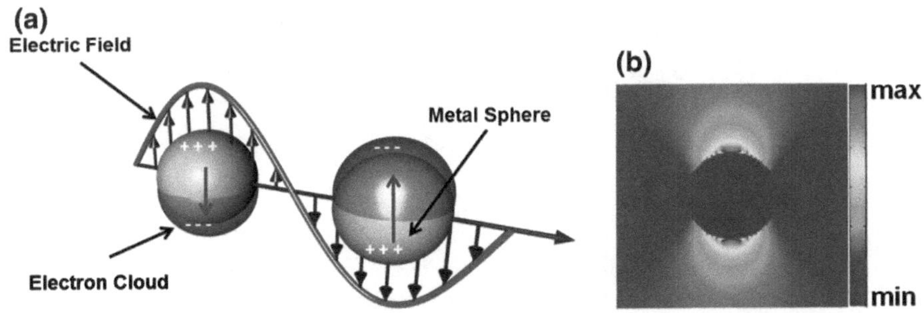

Fig. 1.1 Localized plasmons. **a** Schematic representation of the back and forth movement of the electron cloud around a metal nanoparticle under incident radiation (adapted from Kelly et al. [27]). **b** Electric field intensity map of a 10 nm gold nanosphere with an excitation source propagating along the x-axis and polarized along the y-axis

correlation between the shape of the particle and the SPR frequency related to the distribution of electrons within the metal nanoparticles [35, 36]. For particles with similar dimensions, anisotropic particles with sharp edges generate more intense electric fields compared to their isotropic equivalents (due to higher electron density on the edges). Furthermore, different plasmonic modes can emerge for some particles depending on their shapes [6, 32]. For example, an isotropic spherical nanoparticle shows only one plasmonic mode, whereas an anisotropic metal nanorod shows two plasmonic modes (a transverse mode parallel to its short axis and a longitudinal mode parallel to its long axis). Figure 1.2 summarizes the effect of shape on the electric field intensity (simulated for single particles) and SPR spectra of concave cubes, truncated cubes, and spheres (extinction spectra of nanoparticles were measured in solution for multiple particles) [37].

Plasmonic properties of nanoparticles with various shapes (rod, disk, ring) and dimensions ($\sim$ 30–300 nm) are presented in the following Chaps. (2, 3, 4, and 5) of this thesis through computational and experimental results.

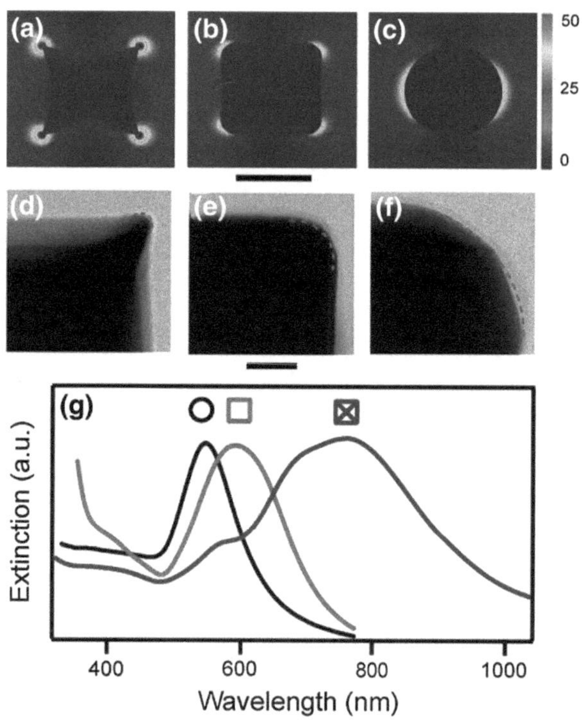

Fig. 1.2 Electric field intensity maps of a **a** concave nanocube, **b** nanocube, and **c** nanosphere with an excitation source propagating in-plane and polarized along x-axis. Scale bar equals to 100 nm. Scanning transmission electron microcope (STEM) images of a corner region on a **d** concave nanocube, **e** nanocube, and **f** nanosphere. The scale bar is 30 nm. The *red dashed line* was used to calculate the radius of curvature of the particles. **g** The absorption spectra of the nanoparticles in water. *Concave cubes* are indicated by crossed *blue squares*, *cubes* as *red squares*, and *spheres* as *black circles* (adapted from Rycenga et al. [37])

1.3 Electrochemical Synthesis of One-Dimensional Structures in Templates

1.3.1 Single and Multisegmented Nanowire Synthesis

There are several synthetic and lithographic approaches for the synthesis of one-dimensional structures on the nanoscale. In particular, the vapor-liquid-solid (VLS) technique has been very successful for the preparation of metallic or semiconducting single and multi-component nanowires on a substrate [38–41]. Although the VLS technique stands out in terms of crystal quality of the nanowires prepared, it offers very limited control over their architecture and composition [23, 42]. In contrast to VLS syntheses, electrochemical deposition within templates offers a direct route to grow multi-segmented metallic and semiconducting nanowires with impressive control over the composition and dimensions of each segment. The electrochemical deposition of metal nanowires in porous templates dates back to 1970 (Possin) [43], in which nanowires of tin, zinc, and other metals in sub-100 nm dimensions were produced in mica templates. Martin, Penner, and Moskovits then introduced the use of porous anodic aluminum oxide (AAO) membranes as a template for metal nanowire synthesis [44–56]. This technique was later expanded by Natan and Keating to make striped nanowires that were exploited as metallic barcodes [57, 58].

A general synthetic scheme describing the electrodeposition of a multisegmented nanowire is presented in Fig. 1.3a. Briefly, a thin film of Ag is evaporated on one side of an AAO template, sealing that side off and serving as a working electrode for the subsequent electrodeposition steps. The template is immersed in an electroplating solution containing ions of a target metal (Au in this scheme) in a conventional three-electrode electrochemical setup. By applying an electric field between the counter and working electrodes, the metal ions are forced to precipitate at the bottom of the pores with the segment length defined by the amount of charge passing through the reference electrode. Following the deposition of the first metal segment, the template is rinsed to avoid cross-contamination of ions. The template is then immersed into a second electroplating solution and potential is applied to electrodeposit the second metal segment (Ni in this scheme). This procedure can be repeated multiple times to create the number of segments desired. Following the electrodeposition steps, nanowires are released into solution by removing the template. Scanning electron microscope (SEM) image (Fig. 1.3b) of a multisegmented Au-Ni nanorod (brighter contrast corresponds to the Au segments) demonstrates the precision in controlling the metal segment length from 5 to 100 nm.

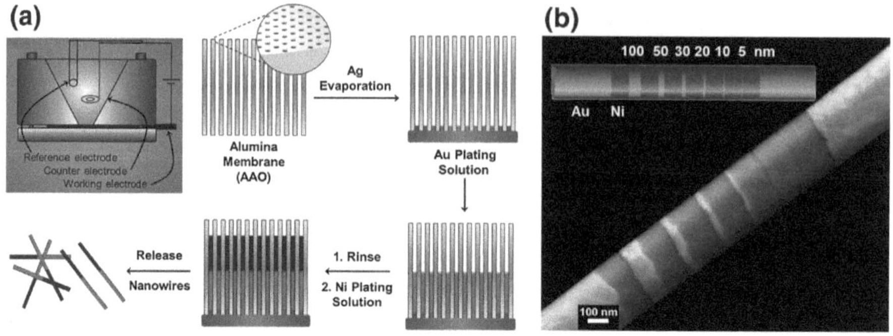

Fig. 1.3 Electrochemical deposition of metal nanowires in porous AAO templates. **a** Synthetic scheme illustrating the fabrication steps of a multisegmented nanowire, **b** SEM image of a Au–Ni multisegemented nanowire with varying segment lengths (adapted from Mirkin et al. [8, 59])

1.3.2 On-Wire Lithography

On-wire lithography (OWL) is a template-based electrochemical process, invented and developed by the Mirkin group, to form one-dimensional solution dispersible arrays of metal nanorods with programmably created nano- and micron-scale gaps on a solid support [8, 9, 11, 19, 59–64]. Synthetic scheme of the on-wire lithography process is presented in Fig. 1.4. During the electrodeposition steps, the nanowire is programmed to have alternating segments of sacrificial (Ni) and target

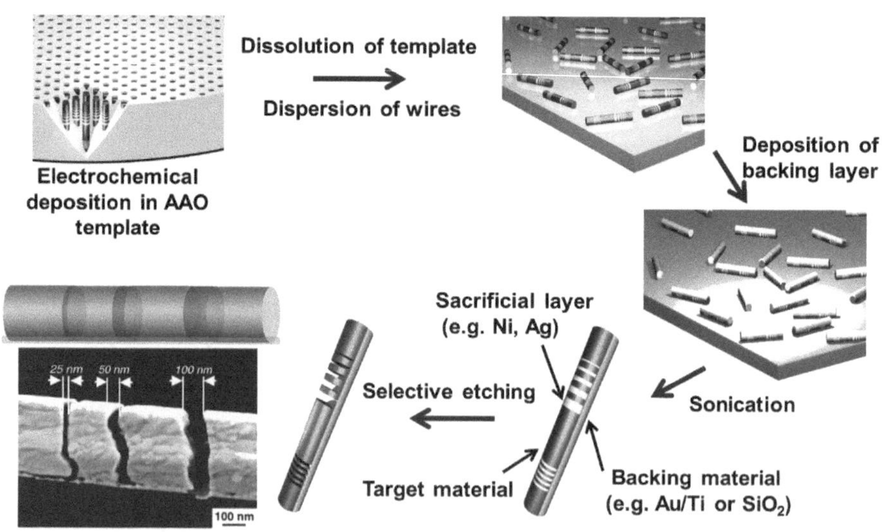

Fig. 1.4 Scheme illustrating the synthetic steps of on-wire lithography along with an STEM image showing a gapped nanorod array prepared following the synthetic steps shown in this scheme (adapted from Qin et al. [8])

materials (Au). Following the electrodeposition steps and the removal of the AAO template, the nanowires are dispersed onto a glass substrate. The key innovation introduced by OWL methodology is the deposition of a very thin silica layer (5–10 nm) on one half of the nanowires. These silica-covered nanowires are released into solution via sonication and the nanowires are treated with an etchant to selectively remove the sacrificial segments without etching the target material. The silica layer serves as a support for the un-etched target materials so that the Au nanorods preserve their original pre-programmed positions with gaps generated in corresponding positions of the Ni segments. The bottom left is an SEM image of a striped Au nanowire after selective etching of the Ni segments (support layer is Au in this sample for imaging purposes).

OWL provides excellent control over the nanowire diameter (from 35 to 360 nm depending on the membrane pore size), segment length (from 6 nm to a few microns), gap (from 1 nm to a few microns), and composition (Au, Ni, Ag, Pt, Fe) of different nanowire components (Fig. 1.5) [8, 9, 19, 61, 65]. The versatility of this novel technique has allowed for the fabrication of a wide variety of structures with emergent and highly functional properties which have led to advances in SERS, plasmonics, plexcitonics and formed the basis for novel molecular electronic, optoelectronic, encoding, biodetection applications [8, 9, 11, 19, 61, 65].

Although templated electrodeposition offers excellent control over size and composition in the axial dimension, it lacks the ability to control and engineer nanostructure architecture and composition along the radial dimension. In the following chapters of this thesis, major breakthroughs in templated electrochemical synthesis of nanostructures are presented. In particular, Chap. 2 details the integration of semiconductors into OWL methodology, which offers a new platform to study plasmon-exciton interactions; Chap. 3 describes the first synthesis of hybrid core-shell nanowires with tunable plasmonic antennas (from visible to near-IR) and

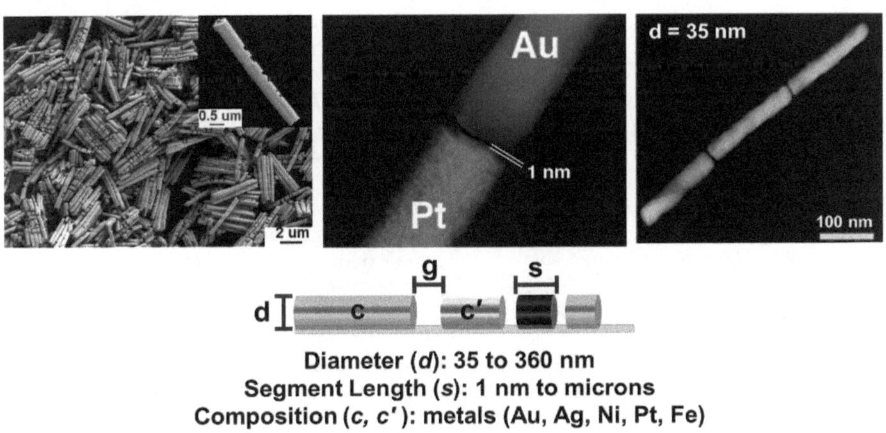

Fig. 1.5 STEM images of gapped nanowires prepared using OWL (*top row*) and a scheme illustrating the geometrical and compositional parameters that can be controlled by OWL (adapted from Mirkin et al. [8, 9, 19, 61, 65])

their use in light harvesting applications; Chap. 4 presents a novel synthetic technique (termed coaxial lithography) which allows one to precisely control dimensions and compositions of a nanowire in both the radial and axial directions; and finally Chap. 5 describes the synthesis of solution dispersible nanorings prepared using coaxial lithography.

References

1. Ozbay, E. (2006). Plasmonics: merging photonics and electronics at nanoscale dimensions. *Science, 311*, 189.
2. Stewart, M. E., Anderton, C. R., Thompson, L. B., Maria, J., Gray, S. K., Rogers, J. A., et al. (2008). Nanostructured plasmonic sensors. *Chemical Reviews, 108*, 494.
3. Maier, S. A. (2007). *Plasmonics: Fundamentals and Applications*. New York: Springer.
4. Atwater, H. A., & Polman, A. (2010). Plasmonics for improved photovoltaic devices. *Nature Materials, 9*(3), 205–213.
5. Ferry, V. E., Munday, J. N., & Atwater, H. A. (2010). Design considerations for plasmonic photovoltaics. *Advanced Materials, 22*, 4794.
6. Jones, M. R., Osberg, K. D., Macfarlane, R. J., Langille, M. R., & Mirkin, C. A. (2011). Templated techniques for the synthesis and assembly of plasmonic nanostructures. *Chemical Reviews, 111*, 3736.
7. Nordlander, P. (2009). The ring: a leitmotif in plasmonics. *ACS Nano, 3*, 488.
8. Qin, L., Park, S., Huang, L., & Mirkin, C. A. (2005). On-wire lithography. *Science, 309*, 113.
9. Osberg, K. D., Schmucker, A. L., Senesi, A. J., & Mirkin, C. A. (2011). One-dimensional nanorod arrays: independent control of composition, length, and interparticle spacing with nanometer precision. *Nano Letters, 11*, 820.
10. Ozel, T., Bourret, G. R., Schmucker, A. L., Brown, K. A., & Mirkin, C. A. (2013). Hybrid semiconductor core-shell nanowires with tunable plasmonic nanoantennas. *Advanced Materials, 25*, 4515.
11. Bourret, G. R., Ozel, T., Blaber, M., Shade, C. M., Schatz, G. C., & Mirkin, C. A. (2013). Long-range plasmophore rulers. *Nano Letters, 13*, 2270.
12. Ozel, T., Nizamoglu, S., Sefunc, M. A., Samarskaya, O., Ozel, I. O., Mutlugun, E., et al. (2011). Anisotropic emission from multilayered plasmon resonator nanocomposites of isotropic semiconductor quantum dots. *ACS Nano, 5*, 1328.
13. Hyun, J. K., & Lauhon, L. J. (2011). Spatially resolved plasmonically enhanced photocurrent from Au nanoparticles on a Si nanowire. *Nano Letters, 11*, 2731.
14. Brittman, S., Gao, H., Garnett, E. C., & Yang, P. (2011). Absorption of light in a single-nanowire silicon solar cell decorated with an octahedral silver nanocrystal. *Nano Letters, 11*, 5189.
15. Knoll, W. (1998). *Annual Review of Physical Chemistry, 49*, 569.
16. Jensen, T. R., Duval, M. L., Kelly, K. L., Lazarides, A. A., Schatz, G. C., & Van Duyne, R. P. (1999). Nanosphere lithography: effect of the external dielectric medium on the surface plasmon resonance spectrum of a periodic array of silver nanoparticles. *The Journal of Physical Chemistry B, 103*, 9846.
17. Coronado, E. A., Encina, E. R., & Stefani, F. D. (2011). Optical properties of metallic nanoparticles: manipulating light, heat and forces at the nanoscale. *Nanoscale, 3*, 4042.
18. Willets, K. A., & Van Duyne, R. P. (2007). Localized surface plasmon resonance spectroscopy and sensing. *Annual Review of Physical Chemistry, 58*, 267.
19. Qin, L., Zou, S., Xue, C., Atkinson, A., Schatz, G. C., & Mirkin, C. A. (2006). Designing, fabricating, and imaging Raman hot spots. *Proceedings of the National Academy of Sciences USA, 103*, 13300.

20. Henzie, J., Lee, M. H., & Odom, T. W. (2007). Multiscale patterning of plasmonic metamaterials. *Nature Nanotechnology, 2*, 549.
21. Linic, S., Christopher, P., & Ingram, D. B. (2011). Plasmonic-metal nanostructures for efficient conversion of solar to chemical energy. *Nature Materials, 10*, 911.
22. Aricò, A. S., Bruce, P., Scrosati, B., Tarascon, J.-M., & Van Schalkwijk, W. (2005). Nanostructured materials for advanced energy conversion and storage devices. *Nature Materials, 4*, 366.
23. Garnett, E. C., Brongersma, M. L., Cui, Y., & McGehee, M. D. (2011). Nanowire solar cells. *Annual Review of Materials Research, 41*, 269.
24. Tang, J., Huo, Z., Brittman, S., Gao, H., & Yang, P. (2011). Solution-processed core-shell nanowires for efficient photovoltaic cells. *Nature Nanotechnology, 6*, 568.
25. Wallentin, J., Anttu, N., Asoli, D., Huffman, M., Åberg, I., Magnusson, M. H., et al. (1057). InP nanowire array solar cells achieving 13.8% efficiency by exceeding the ray optics limit. *Science, 2013*, 339.
26. Ozel, T., Soganci, I. M., Nizamoglu, S., Huyal, I. O., Mutlugun, E., Sapra, S., et al. (2008). Selective enhancement of surface-state emission and simultaneous quenching of interband transition in white-luminophor CdS nanocrystals using localized plasmon coupling. *New Journal of Physics, 10*, 083035.
27. Kelly, K. L., Coronado, E., Zhao, L. L., & Schatz, G. C. (2003). The optical properties of metal nanoparticles: the influence of size, shape, and dielectric environment. *The Journal of Physical Chemistry B, 107*, 668.
28. Anker, J. N., Hall, W. P., Lyandres, O., Shah, N. C., Zhao, J., & Van Duyne, R. P. (2008). Biosensing with plasmonic nanosensors. *Nature Materials, 7*, 442.
29. Arnold, M. D., & Blaber, M. G. (2009). Optical performance and metallic absorption in nanoplasmonic systems. *Optics Express, 17*, 3835.
30. Kreibig, U., & Vollmer, M. (1995). *Optical properties of metal clusters*. Berlin: Springer.
31. Perezjuste, J., Pastorizasantos, I., Lizmarzan, L., & Mulvaney, P. (1870). Gold nanorods: Synthesis, characterization and applications. *Coordination Chemistry Reviews, 2005*, 249.
32. Myroshnychenko, V., Rodriguez-Fernandez, J., Pastoriza-Santos, I., Funston, A. M., Novo, C., Mulvaney, P., et al. (2008). Modelling the optical response of gold nanoparticles. *Chemical Society Reviews, 37*, 1792.
33. Heath, J. R., Knobler, C. M., & Leff, D. V. (1997). Pressure/temperature phase diagrams and superlattices of organically functionalized metal nanocrystal monolayers: the influence of particle size, size distribution, and surface passivant. *The Journal of Physical Chemistry B, 101*, 189.
34. Liz-Marzán, L. M., Giersig, M., & Mulvaney, P. (1996). Synthesis of nanosized gold-silica core-shell particles. *Langmuir, 12*, 4329.
35. El-Sayed, M. A. (2001). Some interesting properties of metals confined in time and nanometer space of different shapes. *Accounts of Chemical Research, 34*, 257.
36. Jin, R., Cao, Y., Mirkin, C. A., Kelly, K. L., Schatz, G. C., & Zheng, J. G. (1901). Photoinduced conversion of silver nanospheres to nanoprisms. *Science, 2001*, 294.
37. Rycenga, M., Langille, M. R., Personick, M. L., Ozel, T., & Mirkin, C. A. (2012). Chemically Isolating Hot Spots on Concave Nanocubes. *Nano Letters, 12*, 6218.
38. Wagner, R. S., & Ellis, W. C. (1964). Vapor-liquid-solid Mechanism of Single Crystal Growth. *Applied Physics Letters, 4*, 89.
39. Wu, Y., Fan, R., & Yang, P. (2002). Block-by-Block Growth of Single-Crystalline Si/SiGe Superlattice Nanowires. *Nano Letters, 2*, 83.
40. Hu, J. T., Odom, T. W., & Lieber, C. M. (1999). Chemistry and physics in one dimension: Synthesis and properties of nanowires and nanotubes. *Accounts of Chemical Research, 32*, 435.
41. Lu, W., & Lieber, C. M. (2007). Nanoelectronics from the bottom up. *Nature Materials, 6*, 841.
42. Tian, B., Zheng, X., Kempa, T. J., Fang, Y., Yu, N., Yu, G., et al. (2007). Coaxial silicon nanowires as solar cells and nanoelectronic power sources. *Nature, 449*, 885.
43. Possin, G. E. (1970). A method for forming very small diameter wires. *Review of Scientific Instruments, 41*, 772.

44. Penner, R. M., & Martin, C. R. (1987). Preparation and electrochemical characterization of ultramicroelectrode ensembles. *Analytical Chemistry, 59*, 2625.
45. Brumlik, C. J., & Martin, C. R. (1991). Template Synthesis of Metal Microtubules. *Journal of the American Chemical Society, 113*, 3174.
46. Martin, C. R. (1991). Template Synthesis of Polymeric and Metal Microtubules. *Advanced Materials, 3*, 457.
47. Brumlik, C. J., Menon, V. P., & Martin, C. R. (1994). Template synthesis of metal microtubule ensembles utilizing chemical, electrochemical, and vacuum deposition techniques. *Journal of Materials Research, 9*, 1174.
48. Martin, C. R. (1994). Nanomaterials: a membrane-based synthetic approach. *Science, 266*, 1961.
49. Martin, C. R. (1995). Template synthesis of electronically conductive polymer nanostructures. *Accounts of Chemical Research, 28*, 61.
50. Martin, C. R. (1996). Membrane-based synthesis of nanomaterials. *Chemistry of Materials, 8*, 1739.
51. Goad, D. G. W., & Moskovits, M. (1978). Colloidal metal in aluminum-oxide. *Journal of Applied Physics, 49*, 2929.
52. Moskovits, M. (1985). Surface-enhanced spectroscopy. *Reviews of Modern Physics, 57*, 783.
53. Preston, C. K., & Moskovits, M. (1993). Optical characterization of anodic aluminum-oxide films containing electrochemically deposited metal particles. 1. Gold in phosphoric-acid anodic aluminum-oxide films. *Journal of Physical Chemistry, 97*, 8495.
54. Almawlawi, D., Liu, C. Z., & Moskovits, M. (1014). Nanowires formed in anodic oxide nanotemplates. *Journal of Materials Research, 1994*, 9.
55. Routkevitch, D., Bigioni, T., Moskovits, M., & Xu, J. M. (1996). Electrochemical fabrication of CdS nanowire arrays in porous anodic aluminum oxide templates. *Journal of Physical Chemistry, 100*, 14037.
56. Martin, B. R., Dermody, D. J., Reiss, B. D., Fang, M. M., Lyon, L. A., Natan, M. J., et al. (1999). Orthogonal self-assembly on colloidal gold-platinum nanorods. *Advanced Materials, 11*, 1021.
57. Nicewarner-Pena, S. R., Freeman, R. G., Reiss, B. D., He, L., Pena, D. J., Walton, I. D., et al. (2001). Submicrometer metallic barcodes. *Science, 294*, 137.
58. Mirkovic, T., Foo, M. L., Arsenault, A. C., Fournier-Bidoz, S., Zacharia, N. S., & Ozin, G. A. (2007). Hinged nanorods made using a chemical approach to flexible nanostructures. *Nature Nanotechnology, 2*, 565.
59. Banholzer, M. J., Qin, L., Millstone, J. E., Osberg, K. D., & Mirkin, C. A. (2009). On-wire lithography: synthesis, encoding and biological applications. *Nature Protocols, 4*, 838.
60. Qin, L., Banholzer, M. J., Millstone, J. E., & Mirkin, C. A. (2007). Nanodisk codes. *Nano Letters, 7*, 3849.
61. Qin, L., Jang, J. W., Huang, L., & Mirkin, C. A. (2007). Sub-5-nm gaps prepared by on-wire lithography: correlating gap size with electrical transport. *Small (Weinheim an der Bergstrasse, Germany), 3*, 86.
62. Qin, L., Banholzer, M. J., Xu, X., Huang, L., & Mirkin, C. A. (2007). Rational design and synthesis of catalytically driven nanorotors. *Journal of the American Chemical Society, 129*, 14870.
63. Banholzer, M. J., Li, S., Ketter, J. B., Rozkiewicz, D. I., Schatz, G. C., & Mirkin, C. A. (2008). An electrochemical approach to and the physical consequences of preparing nanostructures from gold nanorods with smooth ends. *The journal of physical chemistry C, Nanomaterials and interfaces, 112*, 15729.
64. Banholzer, M. J., Osberg, K. D., Li, S., Mangelson, B. F., Schatz, G. C., & Mirkin, C. A. (2010). Silver-based nanodisk codes. *ACS Nano, 4*, 5446.
65. Braunschweig, A. B., Schmucker, A. L., Wei, W. D., & Mirkin, C. A. (2010). Nanostructures enabled by on-wire lithography (OWL). *Chemical Physics Letters, 486*, 89.

Chapter 2
1D Nanowire Synthesis: Extending the OWL Toolbox with Semiconductors to Explore Plasmon-Exciton Interactions in the Form of Long-Range Optical Nanoscale Rulers

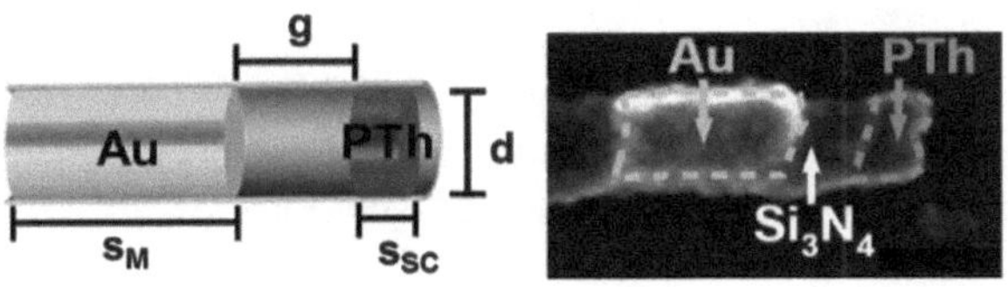

2.1 Introduction

Förster resonant energy transfer (FRET) between molecular fluorophores has been commonly used to measure the distance between molecular entities within biological systems [1, 2]. The technique, however, is limited to the sub-10 nm range [1, 2]. To access longer distances, "rulers" based upon plasmonic architectures have been developed. Indeed, techniques that focus on surface-enhanced Raman-scattering (SERS, surface-to-molecule, sub-20 nm) [3], nanoparticle-based surface energy transfer (surface-to-molecule, sub-50 nm) [4–6] photoluminescence (PL) lifetime (surface-to-molecule, sub-50 nm) [6] and optical coupling between plasmonic dimers (particle-to-particle, sub-100 nm) [7–13] have all been developed for measuring distances on the nanoscale. These structures are often referred to as optical rulers and have strengths and weaknesses depending upon intended application. They are either limited in terms of distance that can be measured or the requirement of two metal particles.

Portions of this chapter have been published in ***Nano Letters*** 13 (5), 2270 (2013).

Co-authors of this work: Gilles R. Bourret, Martin Blaber, Chad M. Shade, George C. Schatz, Chad A. Mirkin.

T. Ozel, *Coaxial Lithography*, Springer Theses,
DOI 10.1007/978-3-319-45414-6_2

The structures that utilize plasmonic particle-fluorophore pairs are quite interesting because they provide a variety of physical parameters that can be influenced by the juxtaposition of these two types of materials [14]. For example, because of their intense and highly localized electromagnetic fields, plasmonic nanostructures can significantly enhance important processes, including photo- and electro-luminescence [15], photoelectric response [16–18] and photocatalytic activity [16, 19–21]. In particular, plasmonic nanoantennas can be used to modulate the rate [22, 23], intensity [24–26] direction [27], polarization [28–32] and spectral shape [33–36] of the spontaneous emission of neighboring luminescent emitters. Since the emission of a strongly-coupled fluorescent/plasmonic pair carries characteristics from both components, such structures are often referred to as plasmophores [14]. Since the plasmon-exciton interaction governing the plasmophore emission is affected by: (i) the overlap of both the absorption and the emission bands of the fluorescent material with the surface plasmon resonance (SPR) of the plasmonic structure, (ii) the near-field intensity generated by the plasmonic construct, and (iii) the distance between the two structures, one can therefore design efficient optical rulers by using specific pairs of plasmonic and luminescent nanostructures [5, 6, 37]. Indeed, the distance-dependent plasmonic modulations of both the PL lifetime and intensity of fluorophores have been measured and used to construct medium-range nanoscale rulers (i.e. sub-50 nm) [5, 6, 37]. However, the effect of distance on the plasmon-induced modification of the PL spectral shape of luminescent emitters has never been studied. Herein, we take advantage of the strong plasmonic fields generated by gold nanorods to develop the first long-range "plasmophore rulers" (valid up to 100 nm) based on the distance-dependent modulation of the emission shape of a luminescent object by a resonant plasmonic nanorod. In addition, we use On-Wire Lithography (OWL) to systematically study the optical responses of plasmonic nanostructures separated by nanoscopic distances from fluorescent semiconducting materials. This geometry, easily accessible by OWL, provides for a test-bed to study the influence of the plasmonic structure on the PL of the neighboring fluorescent materials. A theoretical model of the results is presented which both quantitatively describes the results and shows how plasmon enhancements are responsible for the observed variation of lineshape with gap size, rod length and other structural parameters.

OWL was invented in 2005 for synthesizing one-dimensional arrays of metal nanorods with deliberately designed nano- and micronscale gaps (Fig. 2.1) [38–41]. Recently, we developed a strategy for making structures with diameters as small as 35 nm [42]. This technique involves the sequential electrodeposition of conductive materials within porous anodic aluminum oxide (AAO) membranes to generate striped nanowires with segment lengths ranging from one nanometer to a few microns. OWL can be used to generate solution-processable ordered arrays (as opposed to surface patterning techniques such as e-beam lithography) [43] that are stable in the solid-state and can be made of different materials. The versatility of OWL has allowed for the fabrication of a wide variety of structures that have led to advances in organic electronics [44, 45], SERS, [46–49] plasmonics [42], and biosensing [46–49]. Thus far, OWL has been primarily used to fabricate metal

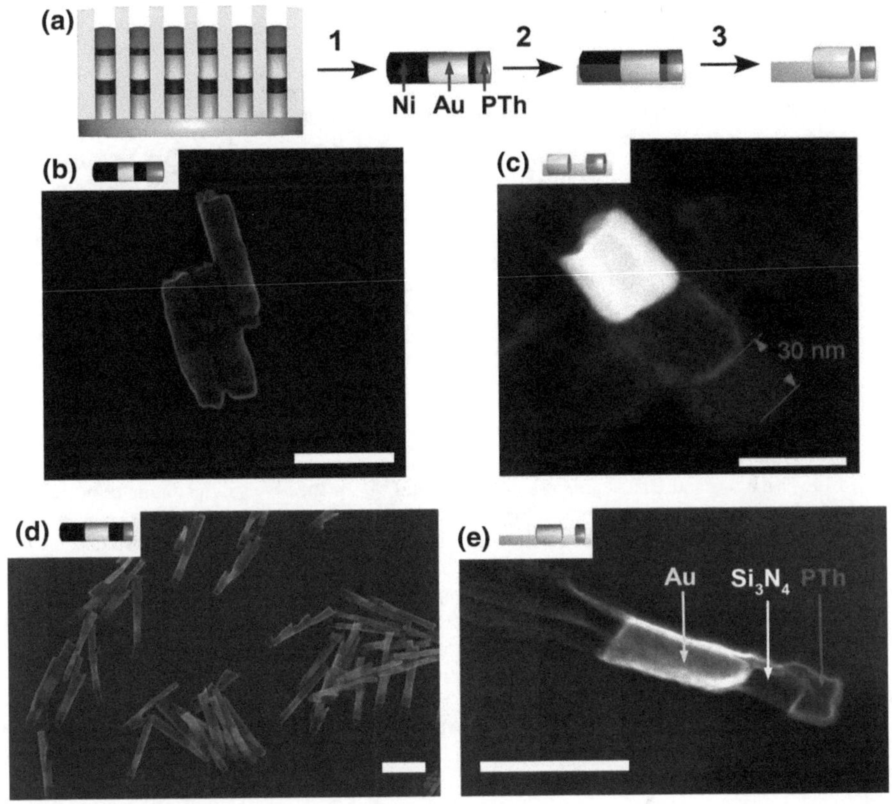

Fig. 2.1 OWL process. **a** (*1*) Release of the multi-segmented nanorods by etching the AAO template and the metal backing. (*2*) Filtration on a porous membrane and sputtering of a thin Si_3N_4 backing layer (*light blue*). (*3*) Chemical etching of the sacrificial layer (Ni, shown in *black*). **b–e** Electron microscopy images of the structures made using OWL. **b** SEM image of Ni-Au-Ni-PTh rods (40 nm diameter) before etching, scale bar: 200 nm. **c** Z-contrast STEM image of the Au-gap-PTh structure made by etching the structure shown in B, scale bar: 50 nm. **d** SEM image of Ni-Au-Ni-PTh rods (70 nm diameter) before etching, scale bar: 400 nm. **e** SEM image of the Au-gap-PTh structure made by etching the structure shown in D, scale bar: 200 nm (Color figure online)

nanostructures (i.e. Pt, Au, Ag and Ni), mostly because they are easily plated from solution and often form stable nanostructures [38, 41, 42, 47].

This chapter describes the first synthesis of plasmonic metal and luminescent organic polymer nanorod dimers based upon OWL, and we explore how such structures can be used to create long range plasmophore rulers. Polythiophene (PTh) was chosen as the luminescent material for its well-defined emission in the visible spectrum [50]. For this study, the OWL method is critical as it allows one to tailor the dimensions of the metal segment and therefore its SPR wavelength (from visible to near-infrared), as well as the gap separating it from the luminescent polymer segment (Fig. 2.1 and Fig. 2.2). This control becomes the basis for the long range plasmophore rulers.

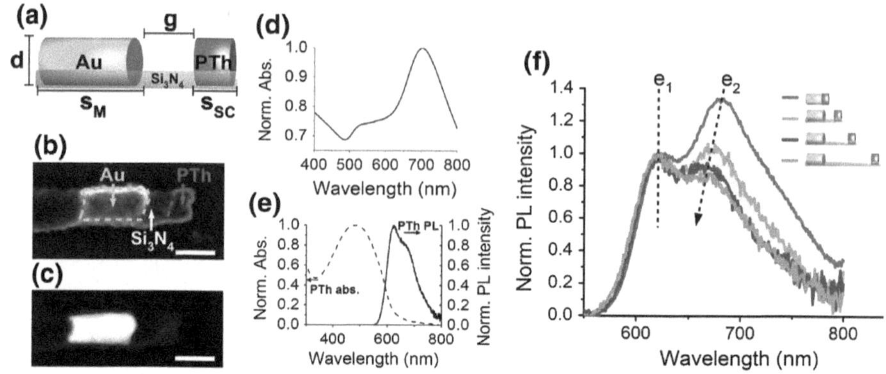

Fig. 2.2 The long-range plasmophore ruler. **a** Schematic depiction of the plasmophore ruler designed via OWL, composed of a gold nanorod and a polythiophene (PTh) nanodisk. The silicon nitride backing acts as a rigid spacer holding the PTh disk at a fixed distance from the Au nanorod. Control over the Au and PTh segment lengths ($\mathbf{s_M}$ and $\mathbf{s_{SC}}$, respectively), the diameter of the rods (**d**) and the gap length (**g**) is demonstrated. (**b**, SEM) and (c, z-contrast STEM) are electron microscopy images of the same hybrid structure composed of a Au nanorod and a PTh disk, whose dimensions (**d** = 40 ± 7 nm, $\mathbf{s_M}$ = 82 ± 5 nm and $\mathbf{s_{SC}}$ = 30 ± 10 nm) have been optimized to maximize the plasmophore ruler signal. The gap length of the structure shown in **b** and **c** is: **g** = 43 ± 12 nm. Scale bars in b and c are 50 nm. **d** Solution-phase normalized absorbance spectrum of the hybrid shown in b and c, with $\lambda_{\mathbf{SPR}}$ = 695 nm. **e** Solution-phase normalized absorbance (*dotted line*, *left axis*) and PL (*solid line*, *right axis*) spectra of the pure PTh nanorods. **f** PL spectra of hybrids excited at 440 nm (same dimensions as the one shown in b and c but with different gap lengths) showing the distance dependence of the plasmonic modulation of the PTh emission (*red curve* no gap, *green curve* **g** = 43 ± 12 nm, *blue curve* **g** = 140 ± 33 nm, *magenta curve* **g** = 330 ± 31 nm) (Color figure online)

2.2 Results and Discussion

The modification of the PTh emission by plasmonic antennas was studied by measuring the solution-averaged PL of AuPTh nanorods (rod diameter: 40 ± 7 nm, PTh disk length: 30 ± 10 nm, no gap) with various gold segment lengths. The broad emission of the pure PTh corresponds to one electronic transition that has a maximum (e_1) at λ_1 = 622 nm and a shoulder (e_2) around λ_2 = 660 nm, and which overlaps well with the SPR of the Au rod (Fig. 2.2e). The Au rod SPR was varied from 600 to 840 nm, corresponding to gold nanorod lengths ranging from 52 to 123 nm (aspect ratios AR = length/diameter varied from 1.27 to 3.24). The position of the Au rod SPR strongly affects the shape of the PTh emission: it modifies the relative intensities of e_1 (I_1) and e_2 (I_2) and significantly shifts the position of λ_2 (Figs. 2.3). To clearly observe this change, all of the PL spectra were normalized at e_1. We use the ratio $R_{2/1} = I_2/I_1$ between the intensities of e_2 and e_1 to quantify the modulation of the PTh PL by the Au antenna. $R_{2/1}$ increases for SPR ranging from 600 to 695 nm, and decreases as the SPR is further red-shifted from the PTh emission band (Fig. 2.4). When the SPR is centered at 695 nm (corresponding to a

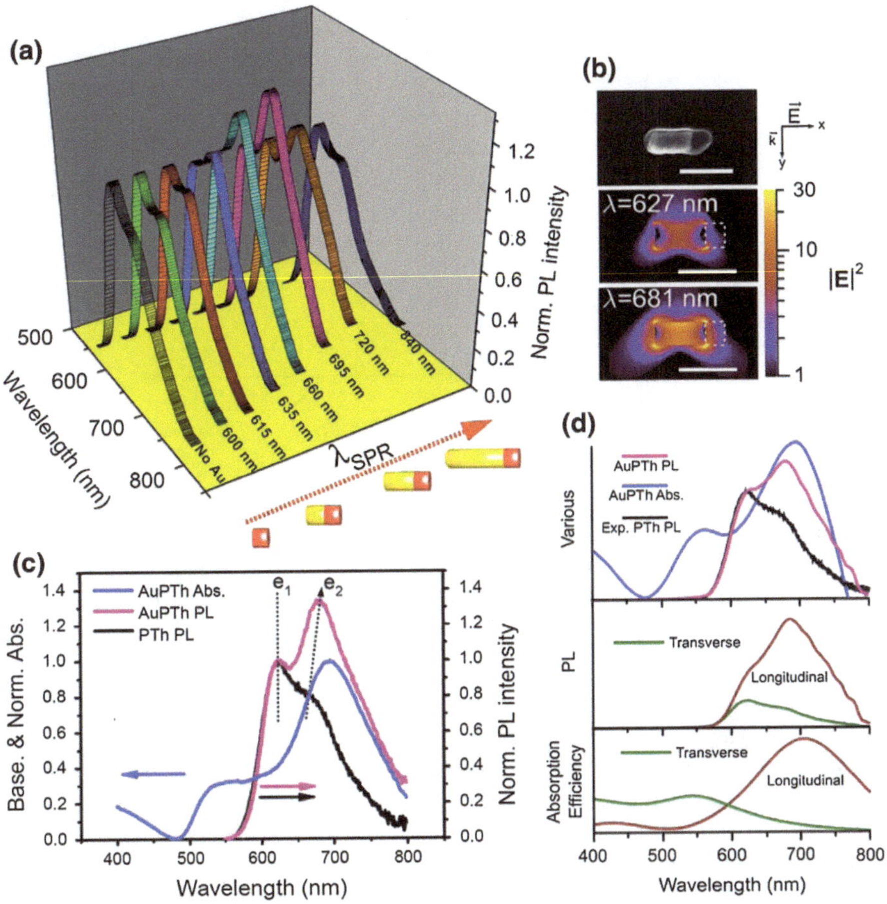

Fig. 2.3 Modification of the polythiophene (PTh) PL by adjacent Au nanorods as a function of the SPR wavelength. **a** PL spectra of Au nanorod/PTh disk structures (no gap, rod diameter: 40 ± 7 nm, PTh length: 30 ± 10 nm) with various gold segment lengths in Nanopure™ water. Excitation wavelength: 440 nm. The curves were normalized to e1 (λ1 = 624 nm). The *black curve* corresponds to the pure PTh emission (no Au rod). As the Au length is increased (*curves* from *left* to *right*), the SPR red shifts (mentioned in *black* next to the PL curves) going from 600 to 840 nm. **b** *Top* SEM image of AuPTh hybrid—SPR centered at 695 nm corresponding to a Au rod length of 82 nm. *Bottom* electric field maps of the simulated AuPTh hybrid (Au rod length: 82 nm and diameter: 44 nm, PTh length: 30 nm), longitudinal polarization (along x axis). Scale bars: 100 nm. The difference in the strength of the electric field inside the PTh (*white dashed lines* in the field maps) between the *center* and *lower panels* indicates the difference in coupling strength between the gold and the PTh and is directly responsible for the increased emission at 681 nm as compared to 627 nm. **c** Overlap between the absorbance (*blue curve*) and the PL spectra (*magenta curve*, excitation wavelength: 440 nm) of a AuPTh hybrid (Au rod length: 82 nm)—SPR centered at 695 nm. The pure PTh PL spectrum is shown as a reference (*black curve*). The curves were base-lined and normalized for clarity. **d** DDA simulations of a AuPTh system similar to the experimental one shown in Fig. 2.3b and c. Length of the gold segment is 82 nm and the diameter is 44 nm. The length of the PTh segment is 30 nm. The simulated spectra (PL and absorbance, top graph) have been obtained by averaging the PL and the absorbance over three different E-field polarizations (along the three main axes)

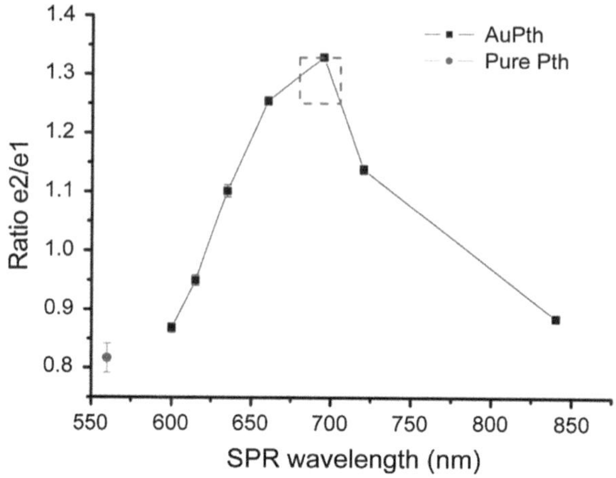

Fig. 2.4 Ratio between the PL intensity at e_1 and e_2 of the AuPTh nanorods as a function of the SPR wavelength. The *red dotted rectangle* highlights the region investigated with the Au/gap/PTh structures. The SPR of these hybrids was centered at 695 ± 15 nm, corresponding to an initial value for AuPTh nanorods (without a gap) of $R_{2/1}$ comprised between 1.33 and 1.25. The y-error bars are due to the uncertainty in determining the ratio e_2/e_1 (Color figure online)

Au rod length of 82 nm), the modification of the emission band shape is the highest. The position of the shoulder e_2 shifts from 660 to 680 nm and $R_{2/1}$ increases by 1.6 fold (going from ≈0.82 without the Au rod to ≈1.33).

The discrete dipole approximation (DDA) was used to model the AuPTh nanorods (based on a reciprocity theorem analysis of the PL spectra where emission enhancement can be directly related to increased radiative rate) [51–53]. The AuPTh hybrid having an experimental $\lambda_{SPR} = 695$ nm (Au rod length: 82 ± 5 nm and diameter: 40 ± 7 nm, PTh length: 30 ± 10 nm) was modeled with a AuPTh structure having a theoretical $\lambda_{SPR} = 695$ nm (Au rod length: 82 nm and diameter: 44 nm, PTh length: 30 nm). The simulated emission of this AuPTh hybrid has $\lambda_1 = 627$ nm and $\lambda_2 = 681$ nm, which are close to the experimental values (i.e. 624 nm and 680 nm, respectively). DDA simulations show that when the Au rod SPR is centered at around 570 nm, the electric field strength integrated over the volume of the PTh is approximately 1.02 times larger at 627 nm than at 681 nm (Fig. 2.5). When the SPR band of the simulated AuPTh rod is centered at 695 nm, it still overlaps with the PTh PL spectra and the electric field at 681 nm is 1.68 times higher than at 627 nm. In this case, $R_{2/1}$ increases by 1.76 fold due to the presence of the gold rod, which is similar to our experimental data (i.e. $R_{2/1}$ increases by 1.6 fold for $\lambda_{SPR} = 695$ nm). This minor discrepancy between the experimental and theoretical peak ratios may be explained by very small differences in quenching between the e_1 and e_2 band. Overall, quenching is expected to be low due to the large spatial extent of the electric field outside the nanoparticle, as quenching is usually only substantive within several nm of the metal surface. Recent experiments on a similar system, poly(3-hexylthiophene), have shown that

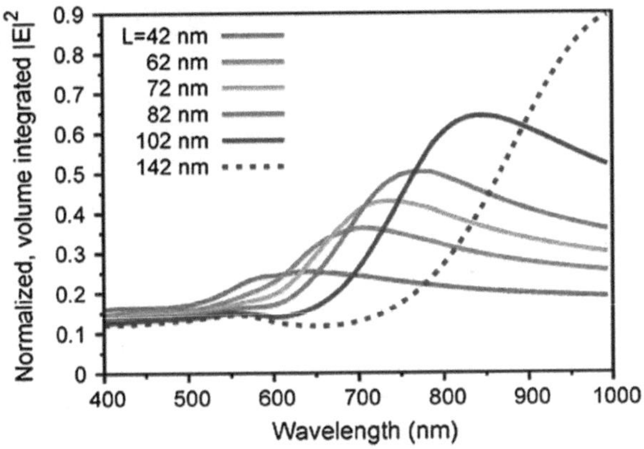

Fig. 2.5 DDA simulation of the effect of increasing the length of the gold segment (diameter: 44 nm) on the volume integrated electric field strength inside the PTh (PTh segment diameter: 44 nm and length: 30 nm). Note that the difference in intensity for the AR = 0.95 (Au length = 42 nm) nanorod (λ_{SPR} = 570 nm) between 627 nm (the approximate location of e_1) and 681 nm (the approximate location of e_2) is negligible, whereas for AR = 1.86 (Au length = 82 nm) case (λ_{SPR} = 695 nm), the volume integrated field strength increases dramatically between e_1 and e_2. When the LSPR shifts past 695 nm, the difference in field strength between e_1 and e_2 reduces and $R_{2/1}$ reduces (see Fig. 2.4)

essentially no quenching occurs at gold-dye distances above 10 nm [29]. As the SPR shifts further to the red, the peak electric field strength increases (Fig. 2.5), but the overlap between the electric field enhancement and the PL spectrum reduces leading to a decrease of $R_{2/1}$. Contours of the local field at λ_1 (627 nm) and λ_2 (681 nm) for simulated structures with a gold rod length of 82 nm (λ_{SPR} = 695 nm) are presented in Fig. 2.3b, while Fig. 2.3d shows modeled DDA PL and absorbance spectra analogous to the experimental results in Fig. 2.3c, as well as the breakdown of the spectra into longitudinal and transverse components. The calculations are in quantitative agreement with experiments. Therefore, the plasmonic modulation of AuPTh nanorod PL depends on: (i) the relative value of I_2 and I_1 (i.e. $R_{2/1}$) without the Au rod and (ii) the difference in the electric field within the PTh segment at λ_2 and λ_1 (directly related to the spectral overlap between the SPR band and the pure PTh PL spectrum). Recent experiments have demonstrated PL emission from pure nanorods that follows the far-field scattering pattern of the longitudinal nanorod SPR upon irradiation of either the transverse SPR, the interband transitions, or the longitudinal SPR itself [54]. However, the quantum yield of the nanorod emission is around 2×10^{-6} which is four orders of magnitude lower than the quantum yield of the emission of polythiophene systems, typically around 10^{-2}–10^{-1} [55]. We did not observe any measurable PL from the pure Au nanorod solutions, and as such the Au nanorod does not contribute significantly to the PL spectrum.

With OWL, we can study the long-range validity of this novel plasmophore ruler by introducing a gap between the Au nanorod and the PTh segment. We use this technique to create billions of highly monodisperse rods and obtain statistically

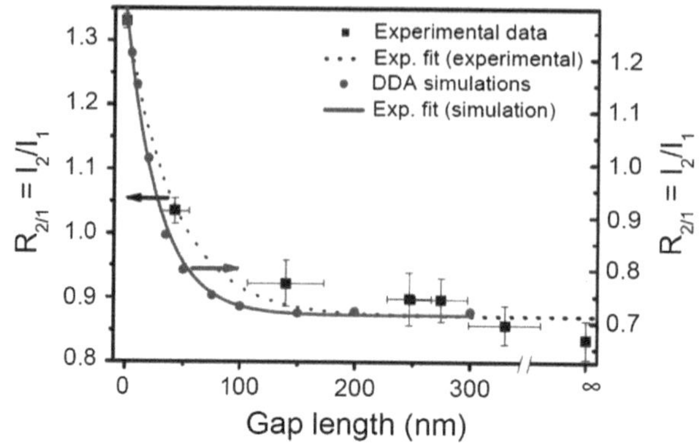

Fig. 2.6 Effect of the gap length on the plasmonic modulation of the polythiophene (PTh) PL. *Black squares* (*left axis*): plot of measured $R_{2/1} = I_2/I_1$ as a function of the gap length, where I_1 and I_2 are the PL intensities at e_1 and e_2, respectively. The Au rod/gap/PTh disk hybrids used for these measurements had a SPR wavelength centered at 695 ± 15 nm. The diameter of the whole structure was 40 ± 7 nm. The Au nanorod was 82 ± 5 nm long and the PTh disk was 30 ± 10 nm long. Each data point corresponds to one solution of nanorods (i.e. one set of structures). The x-error bars are the standard deviations of the gap lengths based on TEM measurements (more than 150 rods were measured for each sample). The y-error bars are due to the uncertainty in determining $R_{2/1}$ from the PL curves. The *dotted black line* is a guide for the eye and corresponds to an exponential decay fit. The point at $x = \infty$ corresponds to the pure PTh case (no Au rod). *Blue circles* (*right axis*): DDA simulations of $R_{2/1}$ for a Au/PTh system having similar absorbance and PL as the experimental one shown in Fig. 2.3c. The Au segment is 82 nm long, and both the gold and the PTh have a diameter of 44 nm. The length of the PTh segment is 30 nm. The *solid blue curve* is a guide to the eyes and corresponds to an exponential decay fit. The data plots (experimental and simulation) are fit to a single-exponential decay of the form $R_{2/1} = a^*\exp(-g/\tau) + b$, where g is the gap length. Experimental data: a = 0.46 ± 0.02, τ = 42.7 ± 6.0, b = 0.87 ± 0.02. DDA simulations: a = 0.58 ± 0.01, τ = 28.4 ± 1.8, b = 0.71 ± 0.01 (Color figure online)

significant data by doing solution-phase measurements that are ensemble-averaged (the sample volume was 500 μL and the concentration of nanostructures was estimated to be in the high picomolar concentration range).

To maximize the plasmonic modulation of the PTh PL, we studied Au nanorod/gap/PTh nanodisk dimers having a SPR centered at 695 ± 15 nm (rod diameter: 40 ± 7 nm, Au length: 82 ± 5 nm, and PTh length: 30 ± 10 nm). The effect of the Au antenna on the PTh emission decreases as the PTh disk is located further away from the Au rod (Fig. 2.2f). This is clearly observed when one examined the dependence of $R_{2/1}$ on the gap size, which can be fitted with a single-exponential decay of the form

$$R_{2/1} = a^*\exp(-g/\tau) + b \tag{2.1}$$

with τ = 42.7 ± 6.0, where g is the gap length (Fig. 2.6, black dotted line). DDA simulations agree well with experimental data (Fig. 2.6, blue circles), with a decay constant τ = 28.4 ± 1.8. A long-range interaction, which becomes negligible above

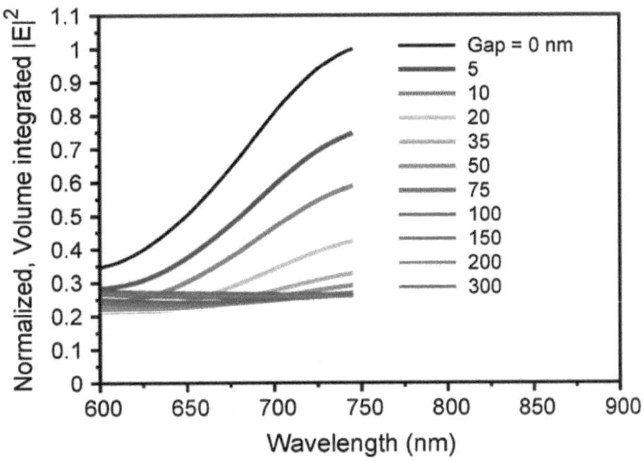

Fig. 2.7 DDA simulation of the effect of increasing the gap between the gold (diameter: 44 nm, length: 82 nm) and the PTh (diameter: 44 nm, length: 30 nm) on the volume integrated electric field strength inside the PTh. There is a 60 % reduction in the field strength when the gap is increased from 0 nm to 20 nm

100 nm, is observed. The electric field at the end of the Au nanorod (longitudinal polarization) decays over a similar length scale (Fig. 2.3b and 2.7).

FDTD simulations performed by Ming et al. suggest that the PL polarization of a fluorophore located at the end of a gold nanorod follows a similar distance dependence (Au rod diameter: 30 nm and length: 70 nm) [31]. This result, coupled with our experimental and theoretical data (Fig. 2.6), suggests that the decay of the electric field intensity away from the Au nanorod accounts for the decrease of $R_{2/1}$ as the gap becomes larger, with nearly 60 % of the reduction in the volume integrated electric field at 695 nm occurring when the gap increases from 0 to 20 nm (Fig. 2.7). By measuring $R_{2/1}$ for a specific fluorophore/plasmonic structure pair, we therefore can determine the distance between the two constructs, similar to the plasmon ruler which provides a measure of the distance between two plasmonic nanostructures [7, 10, 12, 56–58]. Comparison with previous works suggests that the plasmophore ruler presented in this manuscript has a longer range (relative to the nanostructure dimensions) than plasmon rulers based on Au NP dimers [10].

2.3 Conclusions

This work shows how the distance between well-defined plasmonic nanostructures and fluorescent entities can be used to modulate the emission shape of a particular plasmophore. That dependence constitutes a new type of long-range ruler that allows one to probe distances on the 0–100 nm length scale, independently on the signal intensity. A new theory has been developed to describe the results, and it confirms the electromagnetic origin of the dependence of the spectral lineshape on plasmophore structural parameters. It should be noted that these are a proof-of-concept results that provide valuable information for designing more functional rulers, such as the ones used for in vivo studies, where a reference signal cannot practically be incorporated [12].

2.4 Methods and Materials

2.4.1 Materials and Instruments

All chemicals and solutions were used as received. The plating solutions named Cyless, Orotemp 24 Rack and nickel sulfamate were purchased from Technic, Inc. The anodized aluminum oxide (AAO) membranes used for the nanowire synthesis were purchased from Synkera Technologies, Inc. with nominal pore diameters of 35, 55, and 73 nm. The membranes used for filtration were purchased from Whatman (0.2 μm Anodisc 47 membranes). Thiophene (≥99 %), boron trifluoride diethyl etherate and sodium citrate were purchased from Sigma. Nanopure™ water was used.

SEM and z-contrast STEM images were acquired using a Hitachi S-4800 and a Hitachi HD-2300, respectively. Extinction spectra were collected in aqueous solutions using quartz cuvettes (1 cm path length) and a Varian Cary 5000 UV-Vis-NIR spectrophotometer. Fluorescence spectra were collected in aqueous solution using quartz cuvettes (1 cm path length) and a HORIBA Jobin Yvon FluoroLog-3 with excitation and emission slits of 5 mm. Silicon nitride backings were deposited using an AJA Orion Sputter. Electrochemical deposition was done using an EC epsilon potentiostat (BASinc) with a classic three-electrode electrochemical cell. Platinum foil was used as the counter electrode (CE) and the reference electrode was a Ag/AgCl electrode.

2.4.2 Nanowire Synthesis

The nanowires were electrochemically synthesized within AAO membranes (Synkera, Inc.). The general method for producing OWL nanowires made of metal is described in our previous publications [41, 42]. Briefly, 200 nm of silver was thermally evaporated on one side of the AAO templates for electrical connection. Silver was deposited at −940 mV (vs Ag/AgCl) using Cyless plating solution (Technic, Inc.), Au was deposited at −1100 mV (vs Ag/AgCl) using concentrated Orotemp 24 Rack plating solution (Technic, Inc.), and Ni was deposited at −1100 mV (vs Ag/AgCl) using nickel sulfamate plating solution (Technic, Inc.). Prior to the wire synthesis, the pores of the AAO templates were filled with 1000 mC of silver, followed by 250 mC of nickel. These sacrificial segments are necessary to ensure that the nanowire segments of interest are deposited in more uniform and less branched part of the AAO template.

Polythiophene nanorods were grown with a two electrode set-up using a platinum counter electrode (CE), similarly to the synthesis of poly(3-hexylthiophene) nanorods reported by O'Carroll et al. [26, 29]. Prior to the deposition, the cell and the CE were immersed in isopropyl alcohol (IPA), dried under N_2 and under vacuum to remove any residual water. 50 μL of thiophene monomer were dissolved in 5 mL of boron trifluoride diethyl-etherate (BFEE) and deposited with 0.02 s pulses at +3 V, followed by 5 s pulses at −600 mV. The template was successively rinsed with acetonitrile and IPA, and was dried with N_2.

The silver backing was dissolved in a 4:1:1 ethanol: ammonium hydroxide: hydrogen peroxide solution during 20 min. The template was then rinsed thoroughly in H_2O. The AAO template was dissolved in 2 M NaOH (0.5 % citrate by weight) aqueous solution (one hour). The nanowires were recovered using centrifugation (4000 RPM 5 min), and were washed four times with 0.1 % (by weight) citrate aqueous solution. The wires were then dispersed in 24 mL of 0.1 % aqueous citrate solution using sonication before being vacuum filtered onto 47 mm AAO membranes (0.2 μm pores, from Whatman, 2 mL/membrane). A total of twelve membranes were used per nanorod batch in order to avoid aggregation at the surface of the membrane during the filtration process. The membranes were left to dry in air before being sputter coated with a thin layer of silicon nitride (support layer for holding plasmonic Au segment and fluorescent polymer segment after sacrificial Ni segment removal). Radio frequency sputtering plasma was initiated at 40 mTorr pressure and the deposition proceeded at 5 mTorr. The sputtering time was 4 min 15 s, 5 min and 5 min 30 s for nanorods with nominal diameters of 35, 55 and 73 nm respectively. The carrier gas was argon, and the power used was 150 W. To recover the rods coated with Si_3N_4, 20 mL of 3 M NaOH, followed by 100 mL of H_2O, were vacuum filtered through the AAO templates to minimize the interaction between the nanowires and the AAO surface. The nanorods were recovered back into solution by sonicating each of the nanowire-coated AAO membranes in H_2O (10 mL, 0.1 % citrate) for 30 s. 2 mL of 0.5 M NaOH and 0.1 % citrate aqueous solution were then added to the released nanorods in order to dissolve the residual AAO. The solution was left 5 min before being centrifuged at 4000 RPM for 10 min. The rods were then washed with H_2O (0.1 % citrate) and centrifuged at 4000 RPM for 5 min three times. The rods were dispersed in 500 μL of H_2O (0.1 % citrate) and etched for 2 h in 5 % H_3PO_4. Finally they were centrifuged twice and dispersed in 500 μL of H_2O (0.1 % citrate) before any further optical or material characterization.

2.4.3 Discrete Dipole Approximation Simulations

The discrete dipole approximation (DDA) [51, 52] was used to estimate the coupling between the emission of the PTh and the near field of the gold using a model based on the reciprocity theorem [59] in which the time reverse of the PL process is modeled to determine the PTh optical constants that are needed in a DDA calculation of the interaction of a plane wave field with the gold rod/PTh structure. The enhanced fields in the polymer from the DDA calculation are then used to determine the plasmon enhanced PL spectrum using the PTh optical constants.

To implement this theory, the PL spectrum of PTh was modeled as a direct absorption process by relating the emission profile to the imaginary part of the dielectric function

$$\varepsilon(\omega) = \varepsilon_{\mathrm{Re}}(\omega) + i\varepsilon_{\mathrm{Im}}(\omega) \tag{2.2}$$

of an artificial material via

$$\varepsilon_{\mathrm{Im}}(\omega) = \alpha I_{\mathrm{PL}}/\omega^3 \tag{2.3}$$

where I_{PL} is the normalized PL intensity of the PTh, ω is the outgoing photon frequency, and α is an artificial constant that is determined via an empirical fit to the experimental results (here, $\alpha = 3.0\ \mathrm{eV}^3$). The term ω^3 is included to remove the intrinsic frequency dependence of the PL spectrum (which is related to the Einstein "A" coefficient) [60] such that $\varepsilon_{\mathrm{Im}}$ is directly proportional to the square of the transition dipole matrix elements of the emitter.

The real part of the polymer dielectric function is determined by fitting $\varepsilon_{\mathrm{Im}}$ to a sum of oscillators:

$$\varepsilon_{\mathrm{Im}}(\omega) = \mathrm{Im}\left(1 + \sum_i \frac{a}{\omega_i^2 - \omega^2 - i\gamma\omega}\right) \tag{2.4}$$

and taking the real part of the resulting fit (which is a version of a Kramers-Kronig transformation). The gold was modeled using the optical constants of Johnson and Christy [61], with corrections for surface scattering, surface anomalies and polycrystallinity of the OWL structures. The effective electron mean free path was reduced to 10 nm in the simulations [62].

Once the polymer and gold dielectric functions are known, a DDA calculation is used to determine the plasmon-enhanced local field **E** inside the polymer. The plasmon enhanced PL rate is then given by [63]

$$Q_{\mathrm{PL}}(\omega) = \int_{\mathrm{PTh}} \omega^3 \varepsilon_{\mathrm{Im}}^{\mathrm{PTh}} |\mathbf{E}|^2 \mathrm{dV} \tag{2.5}$$

where there is a multiplicative ω^2 (in addition the original ω in the Landau and Lifshitz derivation) to account for the fact that this is a spontaneous emission process. This allows us to relate the rate of energy loss via single particle transitions directly to the creation of photons. Note that the use of $|\mathbf{E}|^2$ associated with a plane wave source in the local field calculation is equivalent to the well-known plane-wave approximation in the SERS enhancement factor calculations [64] and is justified based on a reciprocity theorem analysis as described by Kerker [59]. While the plane wave approximation works best in the quasistatic (small particle) limit, the error will be minor for the particle sizes being considered here due to the rapid decay of $|\mathbf{E}|^2$ within the polymer with distance from the gold surface. Also, it is likely that PL is quenched in the polymer for the first few nm from the gold surface due to electron transfer or other relaxation processes, but this is neglected in the present analysis as a secondary effect as the spatial extent of the enhanced fields is much larger than this quenching region.

References

1. Stryer, L., & Haugland, R. P. (1967). FRET paper. *PNAS, 58*, 719–725.
2. Weiss, S. (1999). Fluorescence spectroscopy of single biomolecules. *Science*, *283*, 1676.
3. Singh, A. K., Khan, S. A., Fan, Z., Demeritte, T., Senapati, D., Kanchanapally, R., et al. (2012). Development of a long-range surface-enhaced raman spectrosocopty ruler. *Journal of the American Chemical Society, 134*, 8662.
4. Yun, C. S., Javier, A., Jennings, T., Fisher, M., Hira, S., Peterson, S., et al. (2005). Nanometal surface energy transfer in optical rulers, breaking the FRET barrier. *Journal of the American Chemical Society, 127*, 3115.
5. Dulkeith, E., Ringler, M., Klar, T. A., & Feldmann, J. (2005). Gold Nanoparticles Quench Fluorescence by Phase Induced Radiative Rate Suppression. *Nano Letters, 5*, 585.
6. Seelig, J., LEslie, K., Renn, A., Kuhn, S., Jacobsen, V., Corput, M. V. D., et al. (2007). Nanoparticle-induced fluorescence lifetime modification as nanoscopic ruler: demonstration at the single molecule level. *Nano Letters, 7*, 685.
7. Reinhard, B. M., Siu, M., Agarwal, H., Alivisatos, A. P., & Liphardt, J. (2005). Calibration of dynamic molecular rulers based on plasmon coupling between gold nanoparticles. *Nano Letters, 5*, 2246.
8. Sonnichsen, C., Reinhard, B. M., Liphardt, J., & Alivisatos, A. P. (2005). A molecular ruler based on plasmon coupling of single gold and silver nanoparticles. *Nature Biotechnology, 23*, 741.
9. Liu, G. L., Yin, Y., Kunchakarra, S., Mukherjee, B., Gerion, D., Jett, S. D., et al. (2006). A nanoplasmonic molecular ruler for measuring nuclease activity and DNA footprinting. *Nature Nanotechnology, 1*, 47.
10. Jain, P. K., Huang, W., & El-Sayed, M. A. (2007). On the universal scaling behavior of the distance decay of plasmon coupling in metal nanoparticle pairs: a plasmon ruler equation. *Nano Letters, 7*, 2080.
11. Reinhard, B. M., Sheikholeslami, S., Mastroianni, A., Alivisatos, A. P., & Liphardt, J. (2007). Use of plasmon coupling to reveal the dynamics of DNA bending and cleavage by single EcoRV restriction enzymes. *PNAS, 104*, 2667.
12. Jun, Y.-W., Sheikholeslami, S., Hostetter, D. R., Tajon, C., Craik, C. S., & Alivisatos, A. P. (2009). Continuous imaging of plasmon rulers in live cells reveals early-stage caspase-3 activation at the single-molecule level. *PNAS, 106*, 17735.
13. Liu, N., Hentschel, M., Weiss, T., Alivisatos, A. P., & Giessen, H. (2011). Three-dimensional plasmon rulers. *Science, 332*, 1407.
14. Lakowicz, J. R., Ray, K., Chowdhury, M., Szmacinski, H., Fu, Y., Zhang, J., et al. (2008). Plasmon-controlled fluorescence: a new paradigm in fluoresecence spectroscopy. *The Analyst, 133*, 1308.
15. Kim, B.-H., Cho, C.-H., Mun, J.-S., Kwon, M.-K., Park, T.-Y., Kim, J. S., et al. (2008). Enhacement of the external quantum efficiency of a silicon quantum dot light-emitting diode by localized surface plasmons. *Advanced Materials, 20*, 3100.
16. Schuller, J. A., Barnard, E. S., Cai, W., Jun, Y. C., & White, J. (2010). Plasmonics for extreme light concentration and manipulation. *Nature Materials, 9*, 193.
17. Ferry, V. E., Munday, J. N., & Atwater, H. A. (2010). Design considerations for plasmonic photovoltaics. *Advanced Materials, 22*, 4794.

18. Mubeen, S., Hernadez-Sosa, G., Moses, D., Lee, J., & Moskovits, M. (2011). Plasmonic photosensitization of a wide band gap semiconductor: converting plasmons to charge carriers. *Nano Letters, 11*, 5548.
19. Linic, S., Christopher, P., & Ingram, D. B. (2011). Plasmonic-metal nanostructures for efficient conversion of solar to chemical energy. *Nature Materials, 10*, 911.
20. Christopher, P., Xin, H., & Linic, S. (2011). Visible-light-enhanced catalytic oxidation reactions on plasmonic silver nanostructures. *Nature Chemistry, 3*, 467.
21. Lee, J., Mubeen, S., Ji, X., Stucky, G. D., & Mokovits, M. (2012). Plasmonic photoanodes for solar water splitting with visible light. *Nano Letters, 12*, 5014.
22. Neretina, S., Qian, W., Dreaden, E. C., & El-Sayed, M. A. (2009). Exciton lifetime tuning by changing the plasmon field orientation with respect to the exciton transition moment direction: CdTe-Au core-shell nanorods. *Nano Letters, 9*, 1242.
23. Wang, Y., Yang, T., Tuominen, M. T., & Achermann, M. (2009). Radiative rate enhancements in ensembles of hybrid metal-semiconductor nanostructures. *Physical Review Letters, 102*, 163001.
24. Pompa, P. P., Martiradonna, L., Torre, A. D., Sala, F. D., Manna, L., Vittorio, M. D., et al. (2006). Metal-enhanced fluorescence of colloidal nanocrystals with nanoscale control. *Nature Nanotechnology, 1*, 126.
25. Singh, M. P., & Strouse, G. F. (2010). Involvement of the LSPR spectral overlap for energy transfer between a dye and Au nanoparticle. *Journal of the American Chemical Society, 132*, 9383.
26. O'Carroll, D. M., Fakonas, J. S., Callahan, D. M., Schierhorn, M., & Atwater, H. A. (2012). Metal-polymer-metal split-dipole nanoantennas. *Advanced Materials, 24*(OP13), 6.
27. Taminiau, T. H., Stefani, F. D., Segerink, F. B., & Hulst, N. F. V. (2008). Optical antennas direct single-molecule emission. *Nature Photonics, 2*, 234.
28. Ming, T., Zhao, L., Yang, Z., Chen, H., Sun, L., Wang, J., et al. (2009). Strong polarization dependence of plasmon-enhanced fluorescence on single gold nanorods. *Nano Letters, 9*, 3896.
29. O'Carroll, D. M., Hofmann, C. E., & Atwater, H. A. (2010). Conjugated polymer/metal nanowire heterostructure plasmonic antennas. *Advanced Materials, 22*, 1223.
30. Ozel, T., Nizamoglu, S., Sefunc, M. A., Samarskaya, O., Ozel, I. O., Mutlugun, E., et al. (2011). Anisotropic emission from multilayered plasmon resonator nanocomposites of isotropic semiconductor quantum dots. *ACS Nano, 5*, 1328.
31. Ming, T., Zhao, L., Chen, H., Woo, K. C., Wang, J., & Lin, H.-Q. (2011). Experimental evidence of plasmophores: plasmon-directed polarized emission from gold nanorod-fluorophore hybrid nanostructures. *Nano Letters, 11*, 2296.
32. Cohen-Hoshen, E., Bryant, G. W., Pinkas, I., Sperling, J., & Bar-Joseph, I. (2012). Exciton-plasmon interaction in quantum dot gold nanoparticle structures. *Nano Letters, 12*, 4260.
33. Kottmann, P. K., Martin, O. J. F., Smith, D. R., Schultz, S. (2000). Spectral response of plasmon resonant nanoparticles with a non-regular shape. *Optics Express, 6*, 213.
34. Ringler, M., Schwemer, A., Wunderlich, M., Nichtl, A., Kurzinger, K., Klar, T. A., et al. (2008). Shaping emission of fluorescent molecules with single plasmonic nanoresonators. *Physical Review Letters, 100*, 203002.
35. Ozel, T., Soganci, I. M., Nizamoglu, S., Huyal, I. O., Mutlugun, E., Sapra, S., et al. (2008). Selective enhancement of surface-state emission and simultaneous quenching of interband transition in white-luminophor CdS nanocrystals using localized plasmon coupling. *New Journal of Physics, 10*, 083035.
36. Zhao, L., Ming, T., Chen, H., Liang, Y., & Wang, J. (2011). Plasmon-induced modulation of the emission spectra of the fluorescent molecules near gold nanorods. *Nanoscale, 3*, 3849.
37. Yun, C. S., Javier, A., Jennings, T., Fisher, M., Hira, S., Peterson, S., et al. (2005). Nanometal surface energy transfer in optical rulers, breaking the FRET barrier. *Journal of the American Chemical Society, 127*, 3115.
38. Qin, L., Park, S., Huang, L., & Mirkin, C. A. (2005). On-wire lithography. *Science, 309*, 113.

39. Qin, L., Jang, J. W., Huang, L., & Mirkin, C. A. (2007). Sub-5-nm gaps prepared by on-wire lithography: correlating gap size with electrical transport. *Small (Weinheim an der Bergstrasse, Germany), 3*, 86.
40. Banholzer, M. J., Li, S., Ketter, J. B., Rozkiewicz, D. I., Schatz, G. C., & Mirkin, C. A. (2008). An electrochemical approach to and the physical consequences of preparing nanostructures from gold nanorods with smooth ends. *The journal of physical chemistry. C, Nanomaterials and interfaces, 112*, 15729.
41. Banholzer, M. J., Qin, L., Millstone, J. E., Osberg, K. D., & Mirkin, C. A. (2009). On-wire lithography: synthesis, encoding and biological applications. *Nature Protocols, 4*, 838.
42. Osberg, K. D., Schmucker, A. L., Senesi, A. J., & Mirkin, C. A. (2011). One-dimensional nanorod arrays: independent control of composition, length, and interparticle spacing with nanometer precision. *Nano Letters, 11*, 820.
43. Gunnarsson, L., Rindzevicius, T., Prikulis, J., Kasemo, B., Kall, M., Zou, S., et al. (1079). Confined plasmons in nanofabricated single silver particle pairs: experimental observations of strong interparticle interactions. *The Journal of Physical Chemistry B, 2005*, 109.
44. Chen, X., Jeon, Y.-M., Jang, J.-W., Qin, L., Huo, F., Wei, W., et al. (2008). Plasmonic focusing in rod−sheath heteronanostructures. *Journal of the American Chemical Society, 130*, 8166.
45. Schmucker, A. L., Barin, G., Brown, K. A., Rycenga, M., Coskun, A., Buyukcakir, O., et al. (2012). Electronic and optical vibrational spectroscopy of molecular transport junctions created by on-wire lithography. *Small, 9*, 1900.
46. Lidong, Q., Shengli, Z., Can, X., Ariel, A. C. S. G., & Mirkin, C. A. (2006). Designing, fabricating, and imaging raman hot spots. *PNAS, 103*, 13300.
47. Qin, L., Banholzer, M. J., Millstone, J. E., & Mirkin, C. A. (2007). Nanodisk codes. *Nano Letters, 7*, 3849.
48. Osberg, K. D., Rycenga, M., Bourret, G. R., Brown, K. A., & Mirkin, C. A. (2012). Dispersible surface-enhaced raman scattering nanosheets. *Advanced Materials, 24*, 6065 doi:10.1002/adma.201202845.
49. Osberg, K. D., Rycenga, M., Harris, N., Schmucker, A. L., Langille, M. R., Schatz, G. C., et al. (2012). Dispersible gold nanorod dimers with sub-5nm gaps as local amplifiers for surface-enhanced raman scattering. *Nano Letters, 12*, 3828.
50. Vardeny, Z., Ehrenfreund, E., Brafman, O., Nowak, M., Schaffer, H., Heeger, A., et al. (1986). Photogeneration of confined soliton pairs (bipolarons) in polythiophene. *Physical Review Letters, 56*, 671.
51. Draine, B. T., & Flatau, P. J. (1994). Discrete-dipole approximation for scattering calculations. *Journal of the Optical Society of America A, 11*, 1491.
52. Draine, B. T., & Flatau, P. J. (2012). User guide for the discrete dipole approximation code DDSCAT 7.2. arXiv:1202.3424 *[physics.comp-ph]* 2012.
53. Flatau, P. J., & Draine, B. T. (2012). Fast near field calculations in the discrete dipole approximation for regular rectilinear grids. *Optics Express, 20*, 1247.
54. Fang, Y., Chang, W.-S., Willingham, B., Swanglap, P., Dominguez-Medina, S., & Link, S. (2012). Plasmon emission quantum yield of single gold nanorods as a function of aspect ratio. *ACS Nano, 6*, 7177.
55. Perepichka, I. F., Perepichka, D. F., Meng, H., & Wudl, F. (2005). Light-emitting polythiophenes. *Advanced Materials, 17*, 2281.
56. Sonnichsen, C., Reinhard, B. M., Liphardt, J., & Alivisatos, A. P. (2005). A molecular ruler based on plasmon coupling of single gold and silver nanoparticles. *Nature Biotechnology, 23*, 741.
57. Liu, G. L., Yin, Y., Kunchakarra, S., Mukherjee, B., Gerion, D., Jett, S. D., et al. (2006). A Nanoplasmonic molecular ruler for measuring nuclease activity and DNA footprinting. *Nature Nanotechnology, 1*, 47.
58. Liu, N., Hentschel, M., Weiss, T., Alivisatos, A. P., & Giessen, H. (2011). Three-dimensional plasmon rulers. *Science, 332*, 1407.

59. Kerker, M. (1969). *The scattering of light and other electromagnetic radiation*. New York: Academic Press.
60. Schatz, G. C., & Ratner, M. A. (2002). *Quantum mechanics in chemistry*; Courier Dover Publications, 2002.
61. Johnson, P. B., & Christy, R. W. (1972). Optical constants of the noble metals. *Physical Review B, 6*, 4370.
62. Kreibig, U., & Fragstein, C. v. (1969). The limitation of electron mean free path in small silver particles. *Zeitschrift für Physik, 224*, 307.
63. Landau, L. D., Bell, J., Kearsley, M., Pitaevskii, L., Lifshitz, E., & Sykes, J. (1984). *Electrodynamics of continuous media* (Vol. 8). Amsterdam: Elsevier.
64. Ausman, L. K., & Schatz, G. C. (2009). On the importance of incorporating dipole reradiation in the modeling of surface enhanced Raman scattering from spheres. *The Journal of Chemical Physics, 131*, 084708.

Chapter 3
Hybrid Semiconductor Core-Shell Nanowires with Tunable Plasmonic Nanoantennas

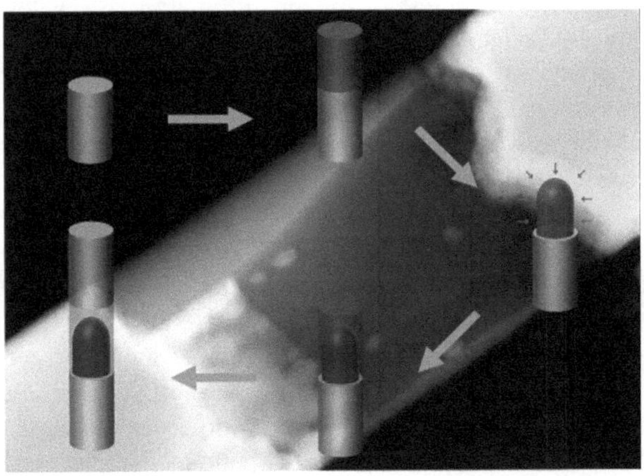

3.1 Introduction

The microstructure of heterogeneous nanowires plays a major role in determining their properties. For example, core-shell nanowires composed of optically active p- and n-type semiconductors have photovoltaic properties that are superior to their planar counterparts due to more efficient charge separation owing to the high interfacial surface area and relatively small distance required to separate electron-hole pairs [1]. Additionally, the size and shape of small metallic structures

---

Portions of this chapter have been published in ***Advanced Materials*** 25, 4515 (2013). Co-authors of this work: Gilles R. Bourret, Abrin L. Schmucker, Keith A. Brown, Chad A. Mirkin. Copyright 2013 WILEY-VCH Verlag GmbH & Co. KGaA, Weinheim.

T. Ozel, *Coaxial Lithography*, Springer Theses,
DOI 10.1007/978-3-319-45414-6_3

determine the spectral location and intensity of the surface plasmon resonance (SPR), an optical effect that when coupled to close proximity of optically active semiconductors, can enhance absorption and emission [2–7]. Despite the far reaching implications of these structural aspects on both the fundamental studies of these systems and their application in fields such as energy harvesting [1, 7], techniques for synthesizing core-shell semiconductor nanowires with robust structural control are very limited. Moreover, including well-defined plasmonically active metallic materials in processes aimed at creating core/shell semiconductor structures is a major challenge.

Methods such as vapor-liquid-solid (VLS) syntheses and cation exchange reactions (solution-phase experiments) have been used to generate core-shell nanowires, but only ones composed of inorganic semiconductors with no ability to couple them to plasmonically active and well-defined metallic nanostructures [8, 9]. In contrast to VLS syntheses, electrochemical deposition into anodic aluminum oxide (AAO) templates offers a direct route to grow multi-segmented metallic and semiconducting nanowires with control over the composition and dimensions of each segment [10]. These methods lay the foundation for techniques, such as on-wire lithography (OWL) [11], which allow one to programmably create gapped and segmented nanowires, structures very useful for studying molecular transport junctions and plasmonics [11–20]. When considering photovoltaic applications, however, current template-controlled electrochemical methods have not been widely used for semiconductor core-shell structures because of the axial nature of materials deposition. In order to study photovoltaic structures of wider potential applicability, one would like to also include low cost, readily available, and chemically tailorable organic materials such as poly(3-hexylthiophene) (P3HT), a material which, despite wide interest owing to its relatively high hole mobility compared to other p-type organic semiconductors [21–26] has seen limited use in nanowire systems due to fabrication challenges [27–30]. The limited combined use of polymer, core-shell, and plasmonic compositions in nanostructured materials highlights the lack of versatility in current synthetic techniques and emphasizes the need for new synthetic techniques to address unmet challenges facing the photovoltaic community.

3.2 Results and Discussion

We report the electrochemical synthesis of solution dispersible, multi-segmented nanowires in which metallic nanostructures are precisely interfaced with core-shell polymeric and inorganic semiconductor materials with nanometer scale resolution (Fig. 3.1). While electrodeposition of metals and inorganic semiconductors within AAO templates produces segments that are mechanically stable, electropolymerized polymers can contract under specific conditions (such as vacuum treatment, pH change, and deposition potential) [31], and leave room for the subsequent growth of a shell around the polymer segment. This phenomenon has been used in the past to

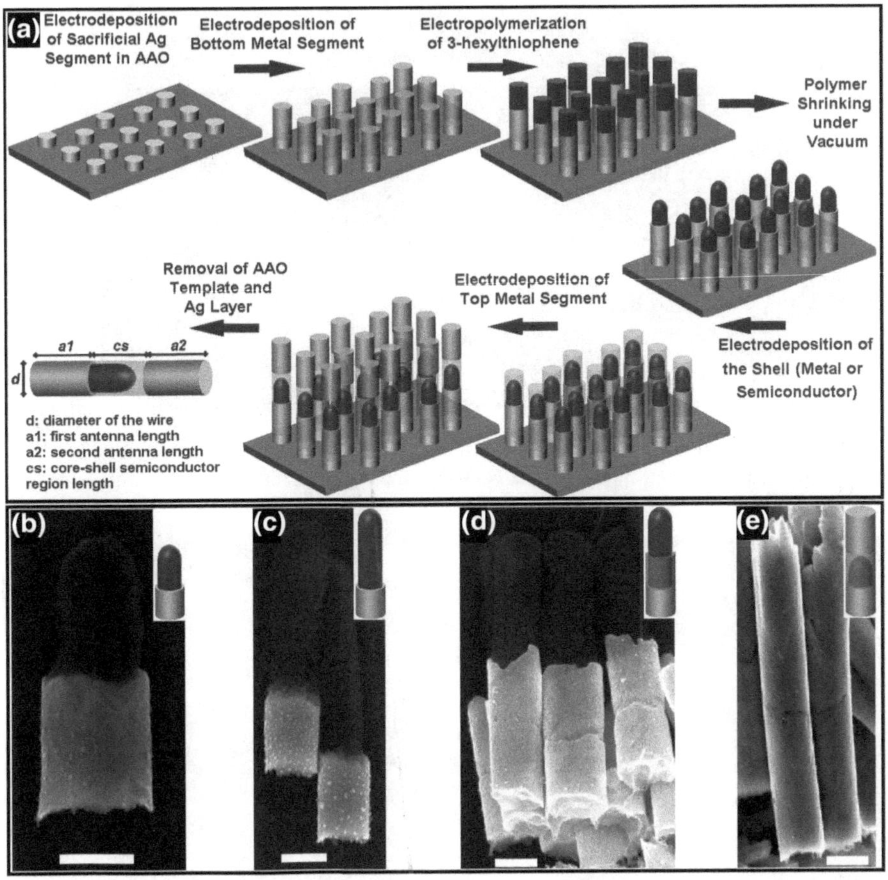

Fig. 3.1 Fabrication of multi-segmented core-shell nanowires. **a** Scheme illustrating the different steps of the fabrication technique. **b**, **c** SEM images of the Au-P3HT nanowires, showing control over P3HT segment length and isotropic shrinking of core material creating room for the shell segment. **d** SEM image of Au shell deposited around the P3HT segment (Au segment-[P3HT core-Au shell]). **e** SEM image of Au shell and Au segment deposited around and on top of the P3HT segment (Au segment-[P3HT core-Au shell]-Au segment). Scale bars are 250 nm

grow different materials around polypyrrole and polythiophene microwires. Depending on the polymer used, the growth of a partial sheath [32] or a complete shell [33] around the polymer core has been reported. For the P3HT system studied here, vacuum treatment was found to induce the isotropic shrinking of the polymer due to the removal of the solvent (Fig. 3.1b–c), allowing for the synthesis of a complete shell of metal and semiconductor around the P3HT core (Fig. 3.1d–e, 3.2, and 3.3). Successive electrodeposition steps were used to integrate metal segments on both ends of the semiconducting segment that can be used as electrical contacts and plasmonic antennas. Our synthetic methodology provides excellent control over the

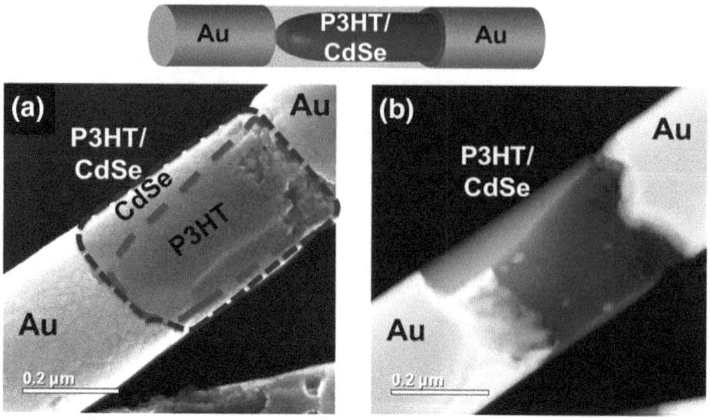

Fig. 3.2 STEM images of Au-[P3HT core-CdSe shell]-Au nanowires (diameter: 280 ± 30 nm) in **a** SE mode, **b** z-contrast mode, indicative of a complete shell formation around the core

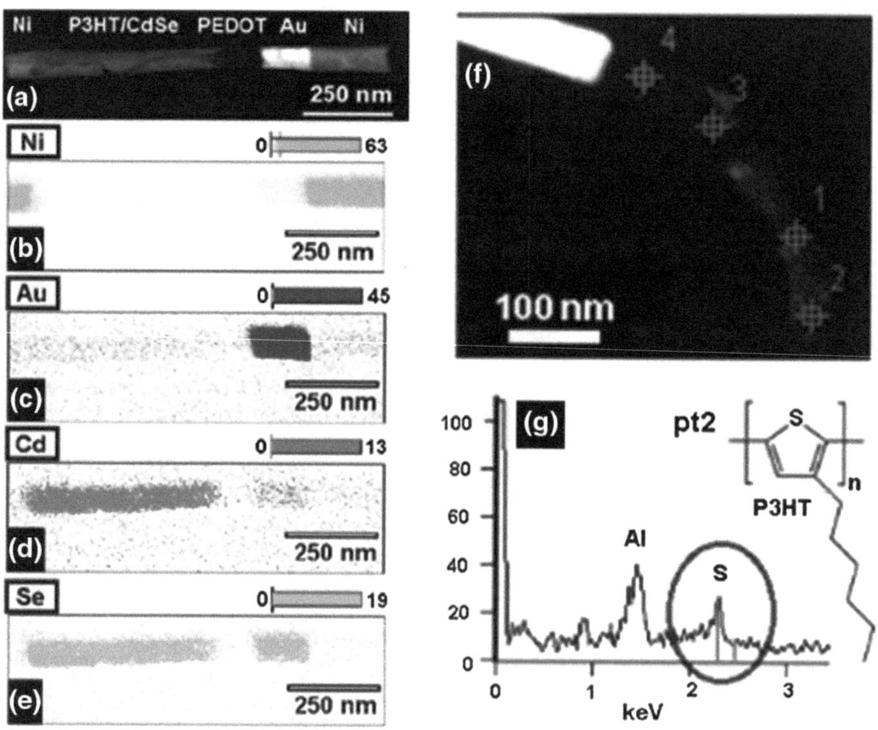

Fig. 3.3 Multisegmented nanowires. **a** STEM image of Au-PEDOT-[P3HT core-CdSe shell]-Ni nanowires (diameter: 60 ± 15 nm) in z-contrast mode and concurrently recorded elemental mapping of the same nanowire for **b** Ni, **c** Au, **d** Cd, **e** Se atoms. Elemental map of Se shows a strong contrast in the position of CdSe shell as expected, but note that there is also some contrast appearing in the location of Au which is due to the overlap of the Au and Se electronic transitions. **f** STEM image of the same nanowire after etching of the shell and top metal segment with 4 points corresponding to elemental scanning points which shows the presence of core material **g** elemental scanning spectrum corresponding to point 2 indicating that the core material is composed of S atoms which proves the presence of P3HT

nanowire diameter (from 30 to 280 nm depending on the membrane pore size), length (from 100 nm to a few microns with a sub-10 nm precision for metals and sub-30 nm precision for the core-shell segment) and composition (Au, Ni, CdSe, PEDOT or P3HT) of the different nanowire components. Fabrication of a photodetector based on a single P3HT/CdSe nanowire with an impressive on/off ratio is reported (comparable to the best reported thin film photodetectors using P3HT/CdSe), [26, 34–36] demonstrating the functionality of the nanowires produced.

In order to evaluate the viability of the approach outlined in Fig. 3.1a, we attempted to synthesize Au-[P3HT-Au] core-shell nanowires. In a typical experiment, porous AAO membranes were coated with a conducting silver backing layer, which was used as the working electrode within a three electrode electrochemical cell. Materials were successively electro-plated or polymerized within the pores of the AAO membrane to generate multi-segmented nanowires. Electrodeposition of the first Au segment under constant potential was followed by the electropolymerization of 3-hexylthiophene (monomer) to create the organic semiconductor core (Fig. 3.1b–c). Since thiophene-based polymers are electropolymerized at high oxidation potentials (E^0_{ox} = 1.7 V vs Ag/AgCl for thiophene) P3HT was grown in boron trifluoride diethyl etherate solution, which is known to lower the oxidation potential of thiophenes [37, 38]. The segment length of P3HT was controlled by deposition time in a cyclical manner (with a precise control over segment length). The current was kept constant during each pulse. Next, the structures were dried under flowing N_2 for 15 s and then held in vacuum for 30 min. Scanning electron microscopy (SEM) images of representative structures formed in this manner show a well-defined gold segment with a polymer segment that has a diameter 15 $\pm$ 5 % smaller than the gold segment diameter (Fig. 3.1b-c). These structures can be subsequently coated with Au to create a cladding on the polymer segment (Fig. 3.1d–e). We hypothesize that Au deposition occurs on the pre-existing metal segment, rather than the exposed polymer, because of the lower electrical resistance of Au compared to the semiconducting polymer segment (electrons prefer the least resistive pathway).

Having shown that core-shell structures can be made with this method, we investigated the possibility of synthesizing core-shell semiconductor heterostructures with relevance to photovoltaic applications. To probe this, we synthesized multi-segmented nanowires with an optically active region composed of a P3HT core and a cadmium selenide shell ([P3HT-CdSe]) using templates with different pore sizes (ranging from 280 to 30 nm). CdSe was grown using the well-established cyclic voltammetry technique [39]. In a typical experiment, wires were grown (see Methods) in a 280 nm pore size template and dispersed on a TEM grid for analysis. Scanning transmission electron microscope (STEM) observation of the same nanowire recorded in different imaging modes suggest the formation of a core-shell segment between the two Au electrodes: a complete segment is observed in the secondary electron (SE) mode image (which gives topographical information of the surface of the wire, Fig. 3.2a) while z-contrast (ZC) mode (which gives compositional information such that the elements with higher atomic numbers give brighter contrast, Fig. 3.2b) provides clear contrast between the CdSe and P3HT sections due to their different electron density, in support of the presence of a core-shell morphology (elemental analysis of the nanowires confirms the core to be P3HT and the shell to be CdSe).

In a subsequent experiment aimed at synthesizing smaller wires using a 60 nm pore size AAO template, we found that the generality of this technique extends to smaller diameter wires and a greater array of materials (Fig. 3.3a). Figure 3.3a shows a multi-segmented nanowire (60 ± 15 nm in diameter) composed of an Au segment followed by a short poly(3,4-ethylenedioxythiophene) (PEDOT) segment (a polymer commonly used in photovoltaics as the hole transport layer) on one side of the [P3HT-CdSe] core-shell active region and a nickel segment on the other side of the active region. The chemical composition of each segment was confirmed via elemental mapping of the nanowire (Fig. 3.3b–e). To further analyze the core-shell structure, the CdSe shell and Ni segment were concurrently etched with nitric acid, and the remaining Au-PEDOT-P3HT segments were imaged and analyzed. The existence of the PEDOT-P3HT polymer segments was confirmed by STEM imaging (Fig. 3.3f), while elemental analysis showed the presence of sulfur in the polymer segments (Fig. 3.3g), consistent with the presence of the PEDOT segment and P3HT core. Moreover, by analyzing the z-contrast STEM images of the nanowires before and after etching of the shell, it is clear that the core-shell segment has a larger diameter (60 nm) than the P3HT core (45 nm).

Optoelectronic characterization of these core-shell nanowires was performed to evaluate their potential utility as optically active structures. Au-[P3HT core-CdSe shell]-Au nanowires (280 nm in diameter) were dispersed on a Si substrate (with a 500 nm thick SiO_2 layer on top) and electrical connection was made between the gold segments of a single wire to preexisting Au contact pads on the substrate using electron-beam lithography followed by electron-beam evaporation of Cr adhesion layer and Au contact layer (Fig. 3.4). Individual wires were then electrically characterized under vacuum using the microscope light as the illumination source.

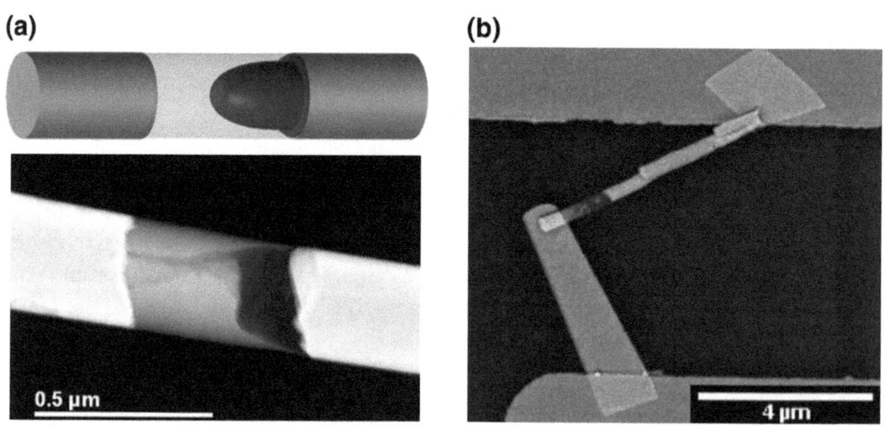

Fig. 3.4 Semiconductor core-shell nanowire. **a** Scheme and STEM image of a p-n radial junction nanowire. **b** SEM image of a fabricated Au-[P3HT core-CdSe shell]-Au nanowire device connected to giant metal electrodes on a Si/SiO_2 chip using e-beam lithography for electrical characterization purposes

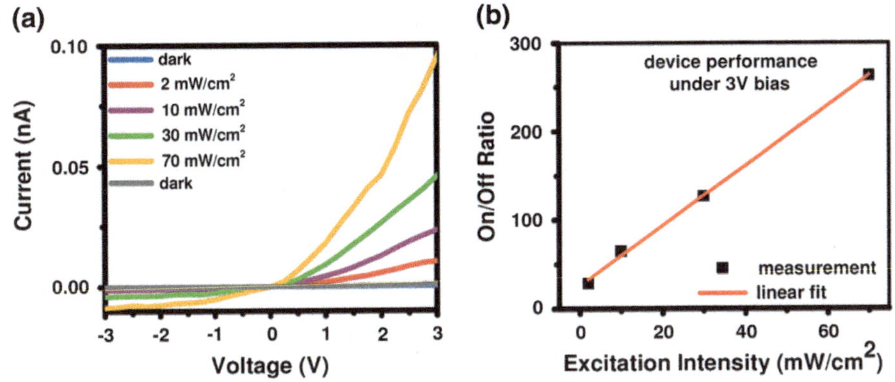

Fig. 3.5 Electrical characterization of semiconductor core-shell nanowire devices. **a** Current-voltage curves demonstrating the photoresponse of the devices with different illumination intensity. The current generated in the hybrid core-shell nanowire drastically increases under light with an on/off ratio of the diode reaching levels up to 280 at a potential difference of 3.75 V. **b** On/off ratios measured for different light intensities under a fixed bias of 3 V. The hybrid nanowire device shows a linear increase in on/off ratio with the light intensity

Current-voltage characterization of a single nanowire exhibited p-n diode behavior with a significant photoconductive effect under illumination (Fig. 3.5a).

The on/off ratio of the conductivity at 3 V potential bias followed a linear trend as a function of the intensity of the white light source (Fig. 3.5b). The hybrid device exhibited a superior on/off ratio performance ($\sim$81) compared to pure CdSe ($\sim$30) and pure P3HT ($\sim$1) devices under the same experimental conditions (30 mW/cm^2 illumination and 1 V potential bias, as depicted in Fig. 3.6a and Fig. 3.6b), clearly showing the synergistic effect of combining the inorganic CdSe shell with the organic P3HT core. This single hybrid nanowire device maintains a remarkable 280 fold on/off ratio under 70 mW/cm^2 excitation at 3.75 V potential bias, which is comparable to what has been previously reported for the best P3HT/CdSe thin film photodetectors [34, 35, 37–39].

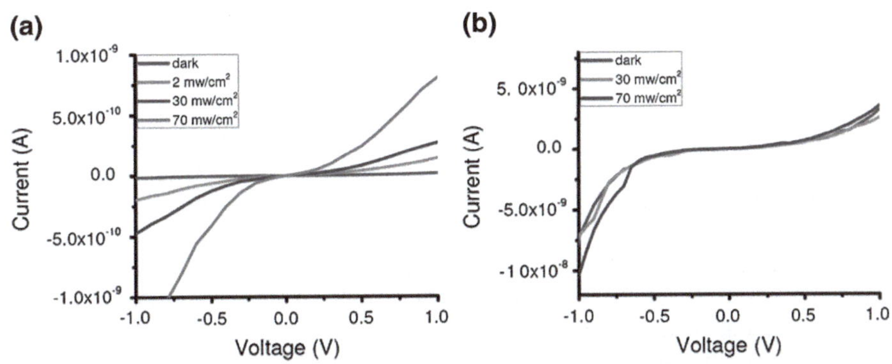

Fig. 3.6 Electrical characterization of semiconductor nanowire devices. Current-voltage curves demonstrating the photoresponse of the devices with different illumination intensity for **a** CdSe and **b** P3HT

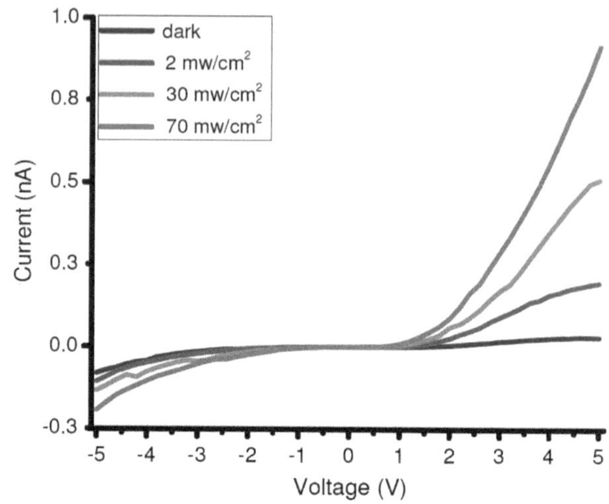

Fig. 3.7 Current-voltage curves of the photoresponse of a Au-[P3HT-CdSe]-Au nanowire where P3HT and CdSe are both connected to both Au electrodes, showing a Schottky diode behavior

One would expect that if the CdSe were also in contact with the top electrode, the device would be symmetric and therefore Schottky diode-type behavior would be observed. Indeed, depending on the extent of shrinking of the P3HT core, Schottky diode behavior was observed in nanowires where the CdSe shell touches both of the Au electrodes (Fig. 3.7).

To explore the possibility of including light-enhancing plasmonic nanostructures in conjunction with core-shell semiconductor heterojunction nanowires, we synthesized [P3HT-CdSe] core-shell nanowires with Au neighboring segments of defined size. In principle, the local electric field enhancement afforded by such plasmonic nanostructures could be used to enhance the exciton generation in the semiconducting region [6, 40, 41]. Multi-segmented nanowires were synthesized with [P3HT-CdSe] core-shell segments located between Au nanorod dimers using templates with different diameters: 30 ± 7 nm (sample **1**), 40 ± 7 nm (sample **2**), and 74 ± 12 nm (sample **3**). Such architectural modifications provide control over the optical absorption properties of the core-shell nanowires (Fig. 3.8a). Specifically, sample **1** exhibited a broad surface plasmon resonance (SPR) band covering ~515–920 nm with λ_{max} ~ 705 nm; sample **2** absorbed over a very broad range of wavelengths (i.e. ~515–1050 nm) with two SPR bands at λ_{max} ~ 670 nm and λ_{max} ~ 865 nm; sample **3** had two distinct SPR bands at λ_{max} ~ 605 nm and λ_{max} ~ 1170 nm, covering the ranges ~530–710 and ~970–1400 nm, respectively.

In order to identify the physical location of the electrical field enhancement afforded by these metal nanostructures, we simulated the localized electric field profiles of these nanostructures (Fig. 3.8b–d) using a commercially available finite-difference time-domain simulation package (Lumerical Inc., Canada). These simulation results highlight the control over the position of the SPR, as well as both the intensity and location of the plasmonic fields around the core-shell segment afforded by this synthetic technique. The dimer geometry allows one to control the position of distinct resonances within the same structure [42]. For example, one can

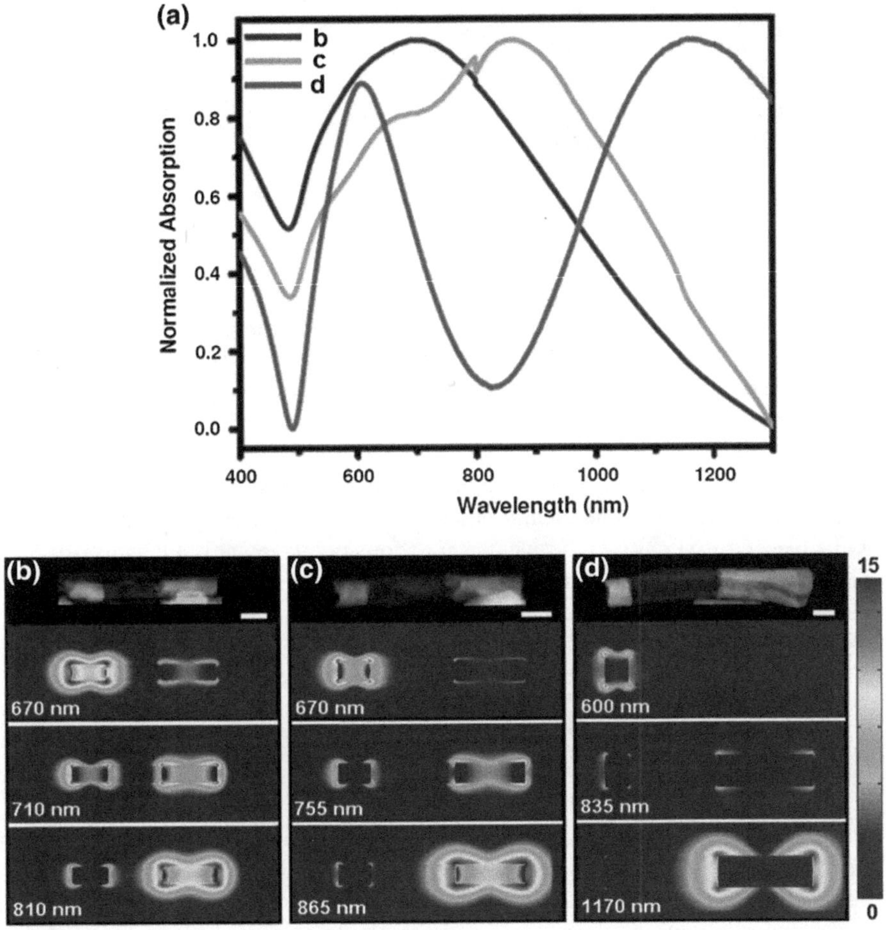

Fig. 3.8 Plasmon coupled semiconductor core-shell nanowires. **a** Extinction spectra of synthesized [P3HT core-CdSe shell] segments (diameters. 30 ± 7 nm (sample **1**), 40 ± 7 nm (sample **2**), and 74 ± 12 nm (sample **3**)) with two Au nanoantennas (with approximate segment length ratio of the first antenna to the second antenna: 2:3, 2:5, and 2:8, respectively). **b–d** Z-contrast STEM images (*top*, note that images are cropped and placed on a *black background* for clarity, scale bars are 50 nm) of the nanowires used for the simulations of the electric-field intensity maps (*bottom*) using a FDTD model. **d** Sample **1,** at wavelengths of 670, 710 and 810 nm. **c** Sample **2,** at wavelengths of 670, 755 and 865 nm. **d** Sample **3,** at wavelengths of 600, 835, 1170 nm

generate very intense fields at specific wavelengths on only one end of the core-shell region (top and bottom E-field maps of Fig. 3.8b–d) or on both ends of the core-shell segment (middle E-field maps of Fig. 3.8b–c). More specifically, one can see that a model based on sample **1** (with the smallest antenna or gold segment length ratio ~2:3) showed strong field generation from the short Au antenna at 670 nm and from the long antenna at 810 nm, while still exhibiting significant field contribution from both gold segments at 710 nm. Sample **3**, due to the large difference in antenna lengths ~2:8, shows strong fields at two distinct wavelengths (~600 and ~1170 nm) with negligible field generated in between the two SPR maxima (at ~835 nm). In the context of photovoltaic or optoelectronic systems, the ability to tune the SPR wavelength of each antenna can be used to enhance the exciton generation over a broad range of wavelengths or at a specific part of the spectrum.

3.3 Conclusions

This work describes a versatile method for efficiently combining core-shell organic and inorganic semiconductor nanowires with plasmonically active structures with sub-100 nm dimensions. Considering that many of the challenges that face the field of nanowire photovoltaics relate to the difficulties associated with electrically contacting the nanowires, the fabrication of high quality semiconductor heterojunctions with built-in metallic leads could dramatically accelerate progress. Finally, the precision with which electrochemically-generated plasmonically-active nanostructures may be placed proximally to these heterojunctions points towards a new test-bed for rapidly studying light-matter interactions in a facile and straightforward manner.

3.4 Methods and Materials

3.4.1 Materials and Instruments

All chemicals and solutions were used as received. The plating solutions named Cyless, Orotemp 24 Rack and nickel sulfamate were purchased from Technic, Inc. 3-hexylthiophene (≥99 %), cadmium sulfate (99 %), selenium dioxide (99.9 %), boron trifluoride diethyl etherate and sodium citrate were purchased from Sigma. Nanopure™ water was used.

Secondary electron (SE) mode and z-contrast STEM images were acquired using a Hitachi HD-2300. Electrochemical deposition was done using an EC epsilon potentiostat (BASinc) with a classic three-electrode electrochemical cell (unless stated otherwise). Platinum mesh was used as the counter electrode (CE) and the reference electrode was a Ag/AgCl electrode.

3.4.2 Nanowire Synthesis

Multisegmented nanowires were synthesized in porous anodized aluminum oxide (AAO) membranes with nominal pore diameters of 280 nm (Whatman Inc.) and of 35, 55 nm (Synkera Technologies Inc.) using a three electrode setup (unless stated otherwise, with a platinum counter electrode and a Ag/AgCl reference electrode) as described in detail in previous publications [11, 17]. For nanowires with diameters of 280 nm, 5000 mC of Au were deposited at −930 mV for the first nanowire segment. P3HT was electrodeposited using 100 μL of 3-hexylthiophene monomer dissolved in 5 mL of boron trifluoride diethyl-etherate (BFEE) solution and polymerized with 0.5 Hz pulses at −150/+150 mA (50 % duty cycle) in a two electrode configuration. Following the polymerization, membranes were rinsed with acetonitrile and IPA, and were dried under 15 s of nitrogen blow and 30 min of vacuum to induce isotropic shrinking of the polymer core. CdSe shell deposition was done using cyclic voltammetry as previously reported (48,000 cycles) [39]. 5000 mC of Au were deposited at −930 mV to form the top Au segment. For nanowires with diameters less than 100 nm, the P3HT deposition was done using a Pt counter electrode and a Pt wire as the reference electrode with 2 μL of 3-hexylthiophene monomer in 5 mL of BFEE solution by cycling between −600 mV and +1000 mV at a rate of 400 mV/s. For sample **1** and sample **2** (35 nm pore diameter), 77 mC (in 11 pulses) of Au were deposited at −1100 mV. 6 cycles of P3HT and 480 cycles of CdSe were used for the semiconducting region. The top Au segment was deposited at −1100 mV for 5 s and 10 s for Sample 1 and 2, respectively. For Sample 3 (55 nm pore diameter), 105 mC (in 15 pulses) of Au, 10 cycles of P3HT, 800 cycles of CdSe, 25 s of Au (at −1150 mV) were used. Nanowires were rinsed in H_2O (0.1 % sodium citrate by weight) by spinning them down 3 times (at 1800 rpm for diameters of 280 nm for 4 min, using the soft acceleration and deceleration mode of an Eppendorf 5417R microcentrifuge), and finally re-suspending them in H_2O (0.1 % sodium citrate by weight). Multisegmented nanowires were produced in high yield with 85 ± 5 % of unbroken nanowires.

As depicted in Fig. 3.9, a rod composed of Au-[P3HT-CdSe]-Au can be observed in secondary electron mode (SE). For better understanding of the core-shell structure, nanowires were investigated under the z-contrast mode, clearly showing the difference in contrast between the core and the shell due to the difference in electron density. Elemental scanning analysis of the active region also supports the presence of a core-shell structure. As can be seen in the point mapping (Fig. 3.9b) in the middle of the active region (point 2), there is sulfur (from P3HT), Cd, and Se peaks showing the contents of both the core and the shell structure. When the point of investigation is located on the outer side of the active region, there are only Cd and Se peaks coming from the CdSe shell.

3.4.3 Electrical Characterization

500 nm oxide coated Si wafers were spin-coated at 500 rpm (10 s) and 4000 rpm (40 s) with a layer of photoresist (S1805, Shipley, USA) and annealed at 115 °C for 1 min. Electrode pads were patterned on the resist with a mask aligner (Süss Microtech MA6 Aligner, Germany) followed by the development of the resist with MF-24A (Microchem, USA) for 1 min. 5 nm Cr and 100 nm of Au were e-beam

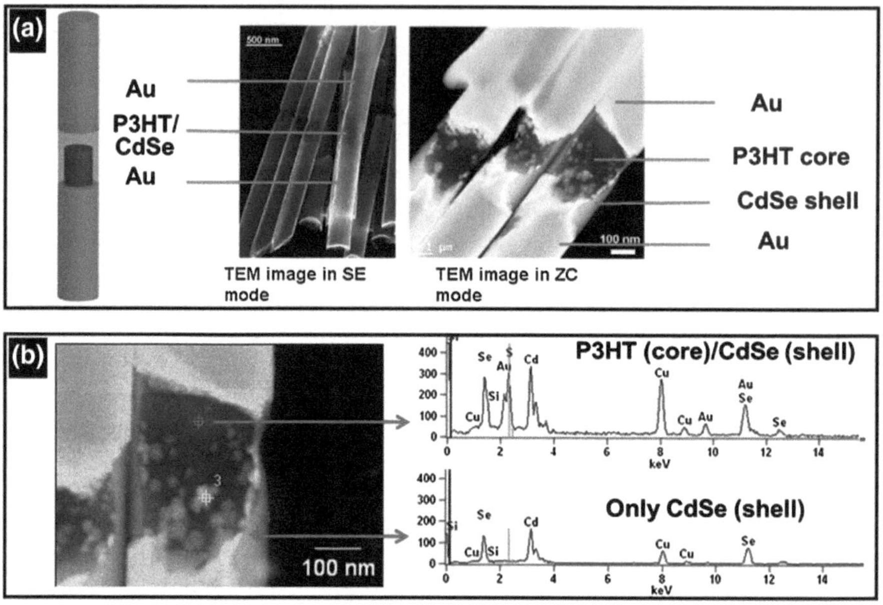

Fig. 3.9 Material characterization of semiconductor core-shell nanowires. STEM images of Au-[P3HT core-CdSe shell]-Au nanowires (diameter: 280 ± 30 nm) in **a** SE mode (*left image*) and zoomed in z-contrast mode (*right image*), indicative of a complete shell formation around the core. **b** STEM image in z-contrast mode and concurrently recorded elemental spectrum of the same nanowire for S, Cd, and Se atoms at two different points. As expected, Cd and Se peaks are intense on the CdSe shell (point 1). The presence of the sulfur peak indicate that the core is composed of P3HT (point 2)

evaporated onto the patterned wafer and wafer was submerged into Remover PG (Microchem, USA) overnight for liftoff to create microelectrode patterned Si chips.

Diluted solutions of multi-segmented nanowires in ethanol were drop-casted on the patterned Si chips. Metal electrode segments of the nanowires were attached to the microelectrodes on the chips using electron beam lithography (EBL). Chips with the nanowires on top were spin-coated with 950 PMMA C7 (Microchem, USA) at 500 rpm (10 s) and 3000 rpm (45 s), and annealed at 180 °C for 2 min. E-beam patterning was done using Quanta FESEM (FEI, USA) with the aid of Nanometer Pattern Generation System (NPGS, JC Nabity Lithography System, Bozeman, MT, USA) at 30 kV acceleration voltage and 24 pA beam current. Patterns were developed with 3:1 IPA/MIBK solution (by volume) for 1 min, and then rinsed with IPA for 5 min and N_2 dried. 5 nm of Cr and 300 nm of Au layers were then e-beam evaporated onto the chips and left in Acetone overnight for liftoff to create electrical connection between the electrode patterns on the Si chips and the individual nanowire segments for testing.

Nanowires were dispersed on 500 nm thick oxide coated silicon wafers containing prefabricated (using photolithography) microelectrode contact pads. Electron beam lithography followed by the e-beam evaporation of 5 nm Cr and 300 nm Au were used to create electrical connection between the Au segments of the individual nanowires and the larger Au contact pads on the chip for electrical testing. Current-voltage characterizations on single nanowires were carried under vacuum using a Keithley 4200-SCS semiconductor characterization system. The built-in white light source of the microscope was used as the illumination source. We measured the output power of the light source with the use of a silicon photodetector from Thorlabs (S130C photodiode power sensor connected to PM200 touch screen power and energy meter console at collection wavelength set to 550 nm).

3.4.4 Finite-Difference Time-Domain Simulations

Electric fields generated by the multi-segmented core-shell nanowires were calculated with the aid of commercial finite-difference-time-domain (FDTD) simulation software developed by Lumerical Solutions Inc., Vancouver, Canada. To simulate the extinction spectra, nanowires were excited by a total field scattered field (TFSF) plane wave source with injection in z-axis with polarization in x-axis at a wavelength range of 350–1300 nm. Pre-defined optical parameters were used directly from the materials library of Lumerical Software for different segments of the nanowires. The refractive index of the medium was set to 1.33 since the wires were suspended in water for the experimental extinction measurements. Electric field simulations were done in 3-dimensions with 0.25 nm mesh size and the electric field intensity maps using 2-dimensional z-axis normal electric field intensity monitors were recorded at different wavelengths of interest between 350 and 1300 nm.

References

1. Garnett, E. C., Brongersma, M. L., Cui, Y., & McGehee, M. D. (2011). Nanowire solar cells. *Annual Review of Materials Research, 41*, 269.
2. Zhu, J., Yu, Z., Burkhard, G. F., Hsu, C. M., Connor, S. T., Xu, Y., et al. (2009). Optical absorption enhancement in amorphous silicon nanowire and nanocone arrays. *Nano Letters, 9*, 279.
3. Muskens, O. L., Rivas, J. G., Algra, R. E., Bakkers, E. P. A. M., & Lagendijk, A. (2008). Design of light scattering in nanowire materials for photovoltaic applications. *Nano Letters, 8*, 2638.
4. Ozel, T., Nizamoglu, S., Sefunc, M. A., Samarskaya, O., Ozel, I. O., Mutlugun, E., et al. (2011). Anisotropic emission from multilayered plasmon resonator nanocomposites of isotropic semiconductor quantum dots. *ACS Nano, 5*, 1328.
5. Ozel, T., Soganci, I. M., Nizamoglu, S., Huyal, I. O., Mutlugun, E., Sapra, S., et al. (2008). Selective enhancement of surface-state emission and simultaneous quenching of interband transition in white-luminophor CdS nanocrystals using localized plasmon coupling. *New Journal of Physics, 10*, 083035.
6. Stuart, H. R., & Hall, D. G. (1998). Island size effects in nanoparticle-enhanced photodetectors. *Applied Physics Letters, 73*, 3815.

7. Ferry, V. E., Munday, J. N., & Atwater, H. A. (2010). Design considerations for plasmonic photovoltaics. *Advanced Materials, 22*, 4794.
8. Tian, B., Zheng, X., Kempa, T. J., Fang, Y., Yu, N., Yu, G., et al. (2007). Coaxial silicon nanowires as solar cells and nanoelectronic power sources. *Nature, 449*, 885.
9. Tang, J., Huo, Z., Brittman, S., Gao, H., & Yang, P. (2011). Solution-processed core-shell nanowires for efficient photovoltaic cells. *Nature Nanotechnology, 6*, 568.
10. Hurst, S. J., Payne, E. K., Qin, L., & Mirkin, C. A. (2006). Multisegmented one-dimensional nanorods prepared by hard-template synthetic methods. *Angewandte Chemie, 45*, 2672.
11. Qin, L., Park, S., Huang, L., & Mirkin, C. A. (2005). On-wire lithography. *Science, 309*, 113.
12. Zheng, G., Qin, L., & Mirkin, C. A. (1938). Spectroscopically enhancing electrical nanotraps. *Angewandte Chemie, 2008*, 47.
13. Chen, X., Jeon, Y.-M., Jang, J.-W., Qin, L., Huo, F., Wei, W., et al. (2008). On-wire lithography-generated molecule-based transport junctions: a new testbed for molecular electronics. *Journal of the American Chemical Society, 130*, 8166.
14. Braunschweig, A. B., Schmucker, A. L., Wei, W. D., & Mirkin, C. A. (2010). Nanostructures enabled by on-wire lithography (OWL). *Chemical Physics Letters, 486*, 89.
15. Bourret, G. R., Ozel, T., Blaber, M., Shade, C. M., Schatz, G. C., & Mirkin, C. A. (2013). Long-range plasmophore rulers. *Nano Letters, 13*, 2270.
16. Osberg, K. D., Schmucker, A. L., Senesi, A. J., & Mirkin, C. A. (2011). One-dimensional nanorod arrays: independent control of composition, length, and interparticle spacing with nanometer precision. *Nano Letters, 11*, 820.
17. Banholzer, M. J., Qin, L., Millstone, J. E., Osberg, K. D., & Mirkin, C. A. (2009). An electrochemical approach to and the physical consequences of preparing nanostructures from gold nanorods with smooth ends. *Nature Protocols, 4*, 838.
18. Schmucker, A. L., Barin, G., Brown, K. A., Rycenga, M., Coskun, A., Buyukcakir, O., Osberg, K. D., Stoddard, J. F., & Mirkin, C. A. (2012). Electronic and optical vibrational spectroscopy of molecular transport junctions created by on-wire lithography. *Small, 9*, 1900.
19. Schmucker, A. L., Harris, N., Banholzer, M. J., Blaber, M. G., Osberg, K. D., Schatz, G. C., et al. (2010). Correlating nanorod structure with experimentally measured and theoretically predicted surface plasmon resonance. *ACS Nano, 4*, 5453.
20. Osberg, K. D., Rycenga, M., Bourret, G. R., Brown, K. A., & Mirkin, C. A. (2012). Dispersible surface-enhaced raman scattering nanosheets. *Advanced Materials, 24*, 6065.
21. Lu, R., Christianson, C., Kirkeminde, A., Ren, S., & Wu, J. (2012). Extraordinary photocurrent harvesting at type-II heterojunction interfaces: toward high detectivity carbon nanotube infrared detectors. *Nano Letters, 12*, 6244.
22. Li, G., Shrotriya, V., Huang, J., Yao, Y., Moriarty, T., Emery, K., et al. (2005). High-efficiency solution processable polymer photovoltaic cells by self-organization of polymer blends. *Nature Materials, 4*, 864.
23. Ma, W., Yang, C., Gong, X., Lee, K., & Heeger, A. J. (2005). Thermally stable, efficient polymer solar cells with nanoscale control of the interpenetrating network morphology. *Advanced Functional Materials, 15*, 1617.
24. Schierhorn, M., Boettcher, S. W., Peet, J. H., Matioli, E., Bazan, G. C., Stucky, G. D., et al. (2010). CdSe nanorods dominate photocurrent of hybrid CdSe−P3HT photovoltaic cell. *ACS Nano, 4*, 6132.
25. Kim, J.-H., Park, J. H., Lee, J. H., Kim, J. S., Sim, M., Shim, C., et al. (2010). Bulk heterojunction solar cells based on preformed polythiophene nanowires via solubility-induced crystallization. *Journal of Materials Chemistry, 20*, 7398.
26. Xu, J., Hu, J., Liu, X., Qiu, X., & Wei, Z. (2009). Stepwise self-assembly of P3HT/CdSe hybrid nanowires with enhanced photoconductivity. *Macromolecular Rapid Communications, 30*, 1419.
27. Huesmann, D., DiCarmine, P. M., & Seferos, D. S. (2011). Template-synthesized nanostructure morphology influenced by building block structure. *Journal of Materials Chemistry, 21*, 408.

28. O'Carroll, D. M., Fakonas, J. S., Callahan, D. M., Schierhorn, M., & Atwater, H. A. (2012). Metal-polymer-metal split-dipole nanoantennas. *Advanced Materials, 24*(OP13), 6.
29. O'Carroll, D. M., Hofmann, C. E., & Atwater, H. A. (2010). Conjugated polymer/metal nanowire heterostructure plasmonic antennas. *Advanced Materials, 22*, 1223.
30. Imahori, H., Kitaura, S., Kira, A., Hayashi, H., Nishi, M., Hirao, K., et al. (2012). A photoconductive, thiophene–fullerene double-cable polymer, nanorod device. *The Journal of Physical Chemistry Letters, 3*, 478.
31. Mukherjee, M., Singh, A., Daillant, J., Menelle, A., & Cousin, F. (1073). Effect of solvent-polymer interaction in swelling dynamics of ultrathin polyacrylamide films: A neutron and X-ray reflectivity study. *Macromolecules, 2007*, 40.
32. Chen, X., Li, S., Xue, C., Banholzer, M. J., Schatz, G. C., & Mirkin, C. A. (2008). Plasmonic focusing in rod− sheath heteronanostructures. *ACS Nano, 3*, 87.
33. Guo, Y., Liu, H., Li, Y., Li, G., Zhao, Y., Song, Y., et al. (2009). *The Journal of Physical Chemistry C, 113*, 12669.
34. Chen, C.-C., Chiu, M.-Y., Sheu, J.-T., & Wei, K.-H. (2008). Photoresponses and memory effects in organic thin film transistors incorporating poly (3-hexylthiophene)/CdSe quantum dots. *Applied Physics Letters, 92*, 143105.
35. Wang, X., Song, W., Liu, B., Chen, G., Chen, D., Zhou, C., et al. (2013). High-Performance Organic-Inorganic Hybrid Photodetectors Based on P3HT: CdSe Nanowire Heterojunctions on Rigid and Flexible Substrates. *Advanced Functional Materials, 23*, 1202.
36. Chiu, M.-Y., Chen, C.-C., Sheu, J.-T., & Wei, K.-H. (2009). An optical programming/electrical erasing memory device: Organic thin film transistors incorporating core/shell CdSe@ ZnSe quantum dots and poly (3-hexylthiophene). *Organic Electronics, 10*, 769.
37. Shi, G., Li, C., & Liang, Y. (1999). High-Strength Conducting Polymers Prepared by Electrochemical Polymerization in Boron Trifluoride Diethyl Etherate Solution. *Advanced Materials, 11*, 1145.
38. Heinze, J., Frontana-Uribe, B. A., & Ludwigs, S. (2010). Electrochemistry of conducting polymers–persistent models and new concepts. *Chemical Reviews, 110*, 4724.
39. Schierhorn, M., Boettcher, S. W., Kraemer, S., Stucky, G. D., & Moskovits, M. (2009). Photoelectrochemical performance of CdSe nanorod arrays grown on a transparent conducting substrate. *Nano Letters, 9*, 3262.
40. Hyun, J. K., & Lauhon, L. J. (2011). Spatially resolved plasmonically enhanced photocurrent from Au nanoparticles on a Si nanowire. *Nano Letters, 11*, 2731.
41. Brittman, S., Gao, H., Garnett, E. C., & Yang, P. (2011). Absorption of light in a single-nanowire silicon solar cell decorated with an octahedral silver nanocrystal. *Nano Letters, 11*, 5189.
42. Mangelson, B. F., Park, D. J., Ku, J. C., Osberg, K. D., Schatz, G. C., & Mirkin, C. A. (2013). Tunable and broadband plasmonic absorption via dispersible nanoantennas with sub-10 nm gaps. *Small (Weinheim an der Bergstrasse, Germany), 9*, 2250.

Chapter 4
2D Nanowire Synthesis: Invention of Coaxial Lithography

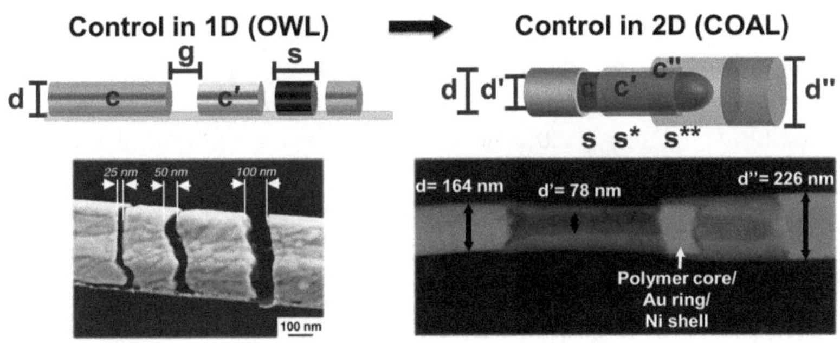

4.1 Introduction

High-resolution lithographic tools are necessary to manipulate and tailor the properties of metals and semiconductors [1–4]. Sustainable energy and optoelectronic device research fields are highly dependent on the development of these technologies [4–9]. For instance, coaxial structures composed of optically active p- and n-type semiconductors can enhance the conversion efficiency of photons into electrical and chemical energy [1, 5–7]. This is due to their large heterojunction areas and energy band bending, which allow for efficient electron-hole separation while minimizing unwanted electron-hole recombination [1]. Similarly, metallic nanostructures in proximity to semiconductors can transfer energy through plasmon resonances to enhance solar light harvesting by the semiconductor [3, 8–10]. Such

---

Portions of this chapter have been published in ***Nature Nanotechnology*** 10, 319 (2015).
Co-authors of this work: Gilles R. Bourret, Chad A. Mirkin.

T. Ozel, *Coaxial Lithography*, Springer Theses,
DOI 10.1007/978-3-319-45414-6_4

structures have allowed researchers to explore novel pathways for energy harvesting and molecular sensing, and they have been proposed as promising entities for improving the conversion of solar light into electrical energy [6, 8, 11]. However, the ability to create semiconductor nanowires with well-defined metallic nanostructures that improve light collection and do not interfere with the electron-hole flow requires precise control over the size and composition of both the core and the shell components. Unfortunately, generating such coaxial nanowire architectures is not readily achievable with current synthetic capabilities.

Methods such as photolithography, electron-beam lithography, dip-pen nanolithography, and nanoimprint lithography, all have been successfully used to prepare functional nanoscale systems on surfaces [3, 12–14]. However, these lithographic systems cannot be used to generate complex nanoscale objects, such as coaxial nanowires. There are a number of methods to coat nanostructures with conformal shells that can be used to generate coaxial nanowires: atomic-layer deposition, sputtering, chemical vapor deposition, solution-phase deposition, and layer-by-layer assembly [2, 5, 7, 11, 15–17]. However, these methods lack the ability to engineer nanostructures along multiple directions (i.e. precise control of the composition, length, diameter, and position of the shell segments), preventing the integration of metal nanorings at specific sites along the nanowires. Such structures could, for example, enhance light emission and absorption within a semiconductor nanowire, by generating an intense electric field localized inside the ring. Moreover, metal nanorings have emerged as important nanoscale materials due to their exceptional optical and magnetic properties [2, 18]. Their synthesis, however, is difficult, making their integration within functional nanowires almost impossible using standard synthetic and fabrication techniques. In contrast to VLS syntheses, electrochemical deposition within anodic aluminum oxide (AAO) templates, pioneered by Martin, Penner, and Moskovits [19–21], and later expanded by Natan and Keating to make striped wires [22, 23], offers a direct route to grow multi-segmented metallic and semiconducting nanowires with impressive control over the composition and dimensions of each segment [10, 13, 19–21, 24, 25]. The use of multi-segmented nanowires to generate nanoscale gaps between metal nanowires by selectively etching a sacrificial segment was developed further by our group and others [13, 25, 26]. The on-wire lithography technique, based on these concepts, was invented and developed by our laboratory to generate one-dimensional arrays of metal nanoparticles and gaps with sub-10 nm resolution [13, 25–27]. However, while this technique allows geometrical control in the axial dimension of the nanowires, it does not provide control in the radial dimension. Such architectural control is essential to synthesize plasmonically and catalytically active metallic nanostructures of the appropriate geometric dimensions in and around coaxial semiconductor nanowires; if realizable, these structures could be used as foundational components for next-generation photovoltaic and photocatalytic systems [6, 8, 15, 28].

4.2 Results and Discussion

This chapter describes a parallel and materials compatible method, termed coaxial lithography (COAL), for synthesizing coaxial nanowires with sub-10 nm resolution in both axial and radial dimensions. COAL allows for the synthesis of multi-compositional coaxial core/shell, core/multi-shell, and asymmetric nanowires via templated electrochemical deposition and selective wet-chemical etching processes (Fig. 4.1). To evaluate the scope of architectural control over the core and shell components, metal nanorings of varying composition (gold, silver, platinum, nickel, and palladium), position, and length (from 8 nm to a few microns) around a wide variety of metal and semiconductor cores (conjugated polymers, metal oxides and metal chalcogenides) of different diameters (from 20 to 400 nm) were synthesized and characterized. Furthermore, the use of COAL to successfully integrate plasmonically active Au nanorings within poly-3-hexylthiophene (P3HT) core/CdSe shell radial junctions is described. Importantly, the plasmonic nanorings

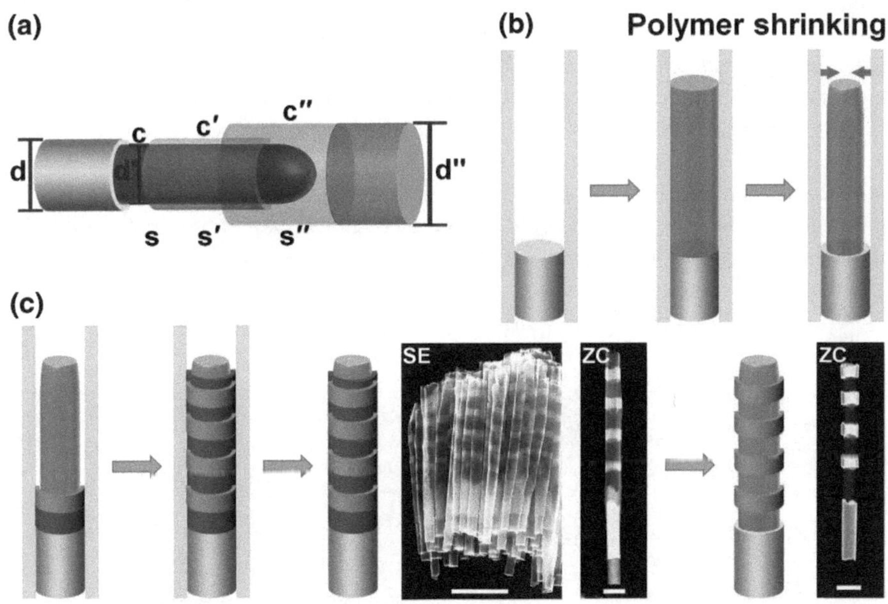

Fig. 4.1 Coaxial lithography. **a** Scheme illustrating the geometrical and compositional parameters that can be controlled by COAL: diameters (d, d′ and d″: from 20 to 400 nm), segment lengths (s, s′ and s″: from 8 nm to few microns), and compositions (polymers: PANI, PPy, PTh, P3HT; metals: Au, Ag, Ni, Pt and Pd; inorganic semiconductors: MnO_2, CdSe and CdS). **b** Scheme illustrating the initial synthetic steps of COAL: electrochemical deposition within the AAO membrane of a metal segment (*yellow*), followed by deposition and shrinking of a polymer segment (*blue*) under vacuum. **c** The following steps for generating metal rings around a polymeric core: deposition of a multi-segmented shell (Au in *yellow* and Ni in *grey*) around the polymer segment, dissolution of the AAO template and etching of the sacrificial shell segment (Ni in this sample). STEM images in SE (secondary electrons) and ZC (z-contrast) modes show typical nanowires before (*left*) and after etching (*right*) the sacrificial Ni shell to generate Au rings (outer diameter: 340 nm) around a polypyrrole core (diameter: 280 nm). Scale bars are 2 μm, 500 nm, and 500 nm, respectively (see color figure online)

do not block the electron-hole flow within these structures and are optically active as shown by the modified photoresponse of the resulting nanowires.

COAL involves the sequential electrodeposition of conductive materials that have different mechanical and chemical stabilities within AAO membranes (Fig. 4.1). Coaxial nanowires are synthesized by initially inducing the radial contraction of electropolymerized polymers within the AAO pores under vacuum, followed by a brief widening step of the AAO pores using 0.5 M NaOH (Fig. 4.1b). This leaves room for the subsequent growth of a shell around the polymer segment [29, 30]. Multi-segmented shells composed of materials with varying reactivities towards wet-chemical etching (such as gold and nickel) are grown around the polymeric core by sequential electrochemical deposition steps (Fig. 4.1c). Following the dissolution of the AAO membrane with aqueous 0.5 M NaOH, subsequent selective etching of the sacrificial segment (nickel) with 50 % by volume aqueous nitric acid generates coaxial nanowires composed of a polymeric core and a striped gold shell [Fig. 4.1c, images were taken in secondary electron (SE) and high-angle annular dark-field imaging z-contrast (ZC) modes of a scanning transmission electron microscope (STEM)]. The dimensions of the negative and positive shell features are deliberately programmed by the thickness of the different segments (nickel and gold), which are electrochemically controlled with nanometer resolution by the amount of charge passing through the electrodes [25].

This approach is compatible with a wide variety of polymeric cores such as polypyrrole (PPy), polyaniline (PANI), polythiophene (PTh), and poly-3-hexylthiophene (P3HT) (Figs. 4.1, 4.2, and 4.3). After sacrificial etching, the dimensions and locations of the gold rings are preserved, generating semiconducting nanowires containing plasmonic Au nanorings with the pre-determined dimensions at the desired locations. To further evaluate the geometric control provided by COAL, we prepared nanowires consisting of a polyaniline core (diameter: 100 nm)

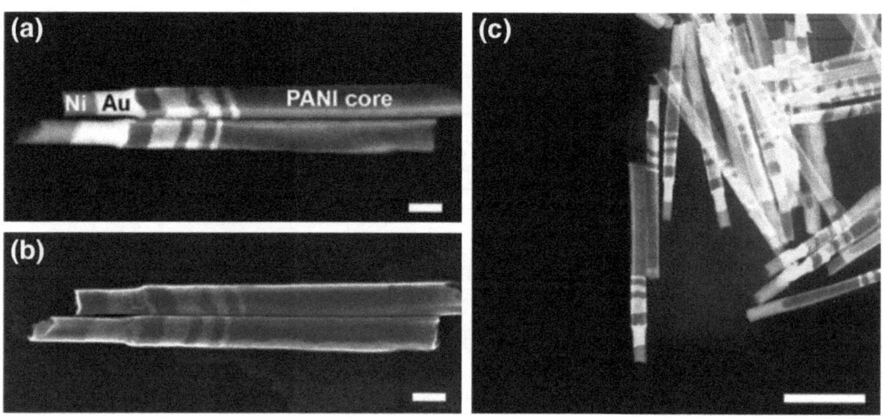

Fig. 4.2 Polyaniline (PANI) core with 3 Au rings of different lengths. **a–c** Electron microscopy images of Ni/Au/PANI core-Ni/Au rings nanowires with Au rings of different lengths (35, 75 and 160 nm). **a** Z-contrast and **b** SE mode images of 2 wires, scale bars: 200 nm. **c** Large scale z-contrast image showing a collection of nanowires, scale bar: 1 μm

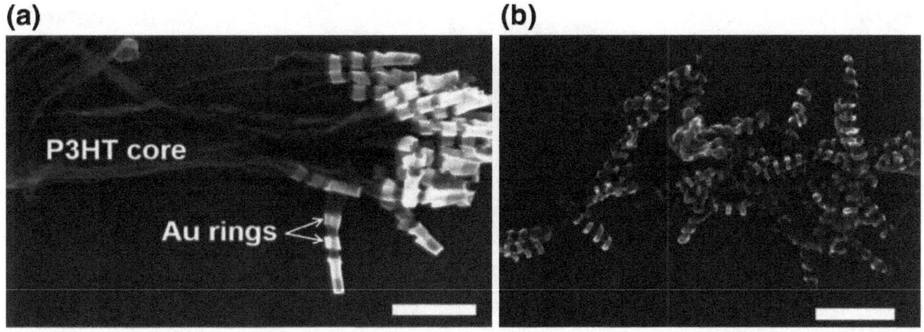

Fig. 4.3 Thiophene derivatives with multiple rings. **a** SEM image of Ni/Au/P3HT core-Au rings nanowires. Scale bar: 1 μm. The length of the Au rings is around 130 nm. **b** SEM image of polythiophene nanowires (40 nm diameter) with four gold rings (outer diameter: 75 nm, inner diameter: 40 nm). Scale bar: 500 nm

with three Au rings (diameter: 140 nm) of different lengths (35, 75, and 160 nm) and characterized them by electron microscopy (Fig. 4.2). The high yield and structural uniformity of the wires generated by COAL are evidenced by large-area electron microscopy images (Fig. 4.1c). Typical standard deviations in segment lengths are 14 % for nickel and 10 % for gold, with more than 80 % of the nanowires having the same number of rings. Size distributions of the inner and outer diameters are typically 15–20 %, mostly due to the dispersity of the AAO template pores. The structure of the gold rings is verified by the dissolution of the polymeric core, which results in the release of intact Au nanorings (this part will be detailed in Chap. 5).

In an attempt to access greater architectural control, we found the diameter of the shell segment can be increased by a pore-widening step, which involves etching of the alumina template with aqueous 0.5 M NaOH (Fig. 4.4). The maximum pore-widening time for reproducible and high yield (~90 %) synthesis of nanorings is 18 min (for 100 nm diameter templates). Figure 4.5 shows how one can

Fig. 4.4 Alternative following steps used to control the shell diameter via pore-widening. This allows for the synthesis of core/shell/shell nanowires, as shown by the STEM images of a PANI core/Au ring/Ni shell nanowire composed of segments that have three different diameters. Scale bar is 250 nm

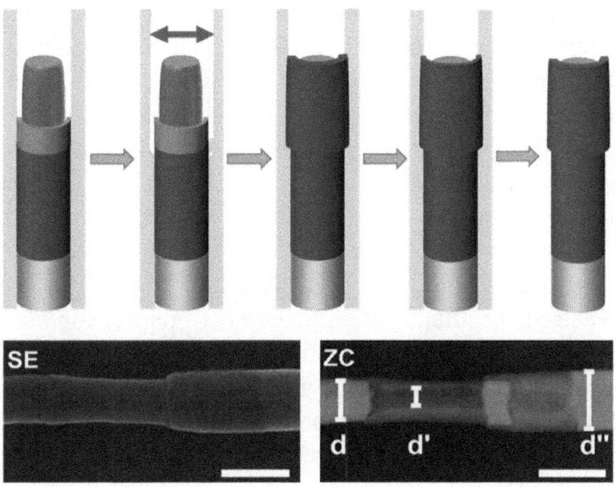

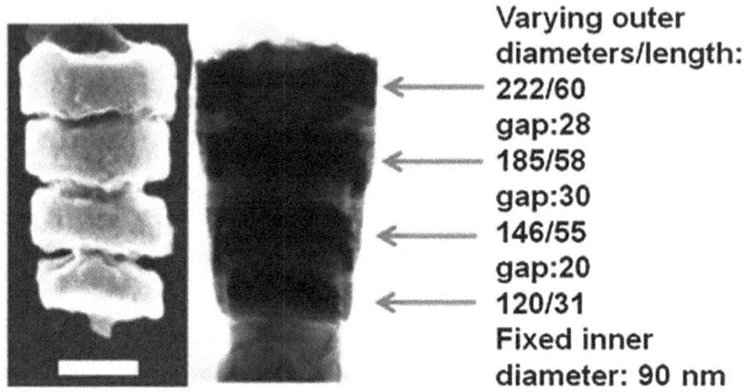

Fig. 4.5 Control over the outer diameter. STEM images of four concentric gold nanorings with increasing diameters around a polymer core without (*left*) and with (*right*) Ni segments. Scale bar equals to 100 nm

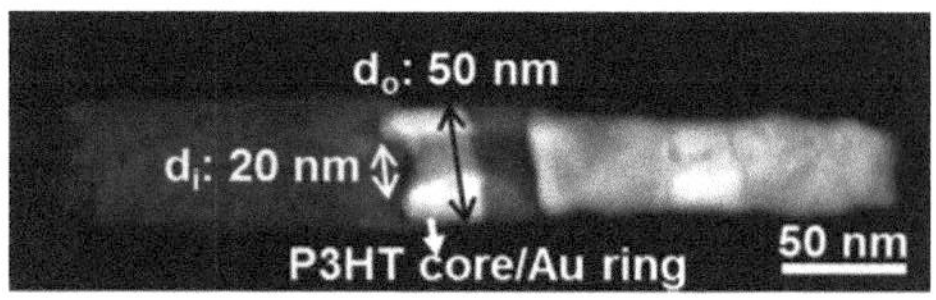

Fig. 4.6 Smallest nanoring. STEM image of a 50 nm Au nanoring around a P3HT core

widen the pores several times on the same template, to generate concentric rings of increasing diameters around a polymer core. Longer exposure times result in the modification of the morphology of the rings due to (partial) merging of the neighboring pores (in this case the products are a mixture of nanorings and non-circular nanoparticles) or complete dissolution of the membrane (after 25 min).

Similarly, control over the inner diameter can be achieved by slowly dissolving the polyaniline core in an ethanol/water mixture. This reduces the polymeric core diameter, and provides significant tuning in terms of inner shell diameter (the final inner diameter can be reduced down to ~65 % of the original diameter). The smallest radial feature currently achievable with COAL is a metal shell having a 20 nm inner diameter and a 50 nm outer diameter (Fig. 4.6). These allow for the synthesis of coaxial nanowires with multiple shells that differ in composition and diameter (Fig. 4.4).

To generate core/shell nanowires with an inorganic core, we developed an alternative approach (Fig. 4.7), which involves initially dissolving the polymeric core (polyaniline) in acetone and then etching the sacrificial segments within the AAO membrane. Removal of the sacrificial metal shell segments prior to the deposition of new core material is important as the subsequent material deposition occurs both at the bottom metal segment and the inner walls of the shell segments leading to a non-uniform deposition. However, in the case of the sacrificial shell removal, the rings remain fixed to the pores of the AAO membrane at their original programmed locations (Fig. 4.7) and they are electrically isolated from the bottom electrode segment. Subsequent electrodeposition occurs at the bottom of the pores, generating a nearly conformal contact between the core and shell segments. This pathway allows for the synthesis of coaxial nanowires with a core composed of practically any material that can be electrodeposited (metals, organic or inorganic semiconductors), making COAL a highly versatile technique.

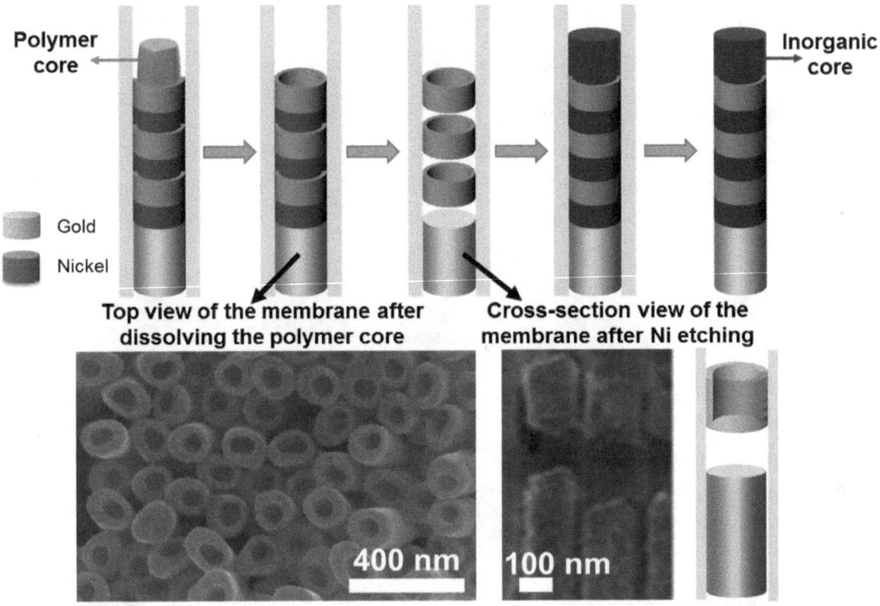

Fig. 4.7 Generalization of COAL to inorganic cores. Scheme illustrating the modified synthetic steps. From *left* to *right* dissolution of the PANI core (shown in *blue*), etching of the sacrificial segments (*grey*) within the AAO template, deposition of the new core material (*purple*), dissolution of the AAO template. *Top-view* SEM image of Au tubes after removal of the PANI core and the AAO membrane (*bottom left*), scale bar: 2 μm. Cross-section SEM image showing that the ring location is preserved inside the membrane after dissolution of the PANI core and etching of the Ni sacrificial rings (*bottom right*). Scale bar: 100 nm (see color figure online)

To test the materials generality of core deposition, we synthesized several structures composed of different inorganic cores (MnO_2, CdS, and Ni) with well-defined metal nanorings. Elemental mapping via energy-dispersive X-ray spectroscopy (EDS) confirms the formation of chemically pure rings around each of the nanowire cores. In addition, we tested the materials generality of the rings located around an inorganic nanowire (Fig. 4.8). As a proof-of-concept, we synthesized nanowires with four distinct rings, composed of Au, Pt, Ag and Pd.

Attempts to determine resolution show that COAL provides sub-10 nm control over shell length, with the smallest structure made to date being a Pt ring with an 8 nm length surrounding an 80 nm diameter MnO_2 wire (Fig. 4.9).

Finally, to push the limits of the structural tailorability possible via COAL, we attempted to synthesize a nanoring-embedded hybrid core/shell semiconductor nanowire. This target architecture consists of a p-type core surrounded with a gold ring at a deliberately programmed location. This entire structure is surrounded by an n-type shell with pure gold segments addressing opposite ends. This novel nanowire architecture was not only chosen for its complexity but also because it should enhance the absorption of visible light without significantly blocking the electron-hole flow. COAL was used to make the structure, which involved initially

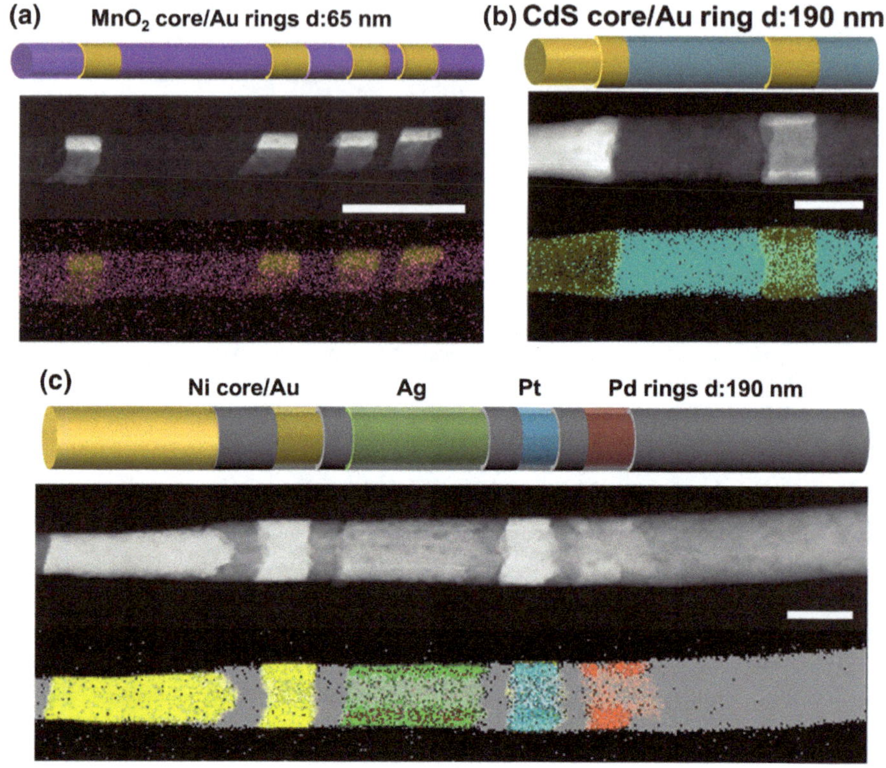

Fig. 4.8 ZC STEM images and elemental map of: **a** multiple Au rings (inner diameter: 55 nm) around an MnO_2 core (MnO_2 diameter: 65 nm; elemental map: *yellow* corresponds to Au and *purple* to Mn), **b** single Au ring (inner diameter: 140 nm) around a CdS core (CdS diameter: 190 nm; elemental map: *yellow* corresponds to gold and *blue* corresponds to Cd), and **c** rings composed of Au, Ag, Pt, Pd (inner ring diameters: 115 nm) around a Ni core (Ni diameter: 190 nm; elemental map: *grey* corresponds to Ni, *yellow* to Au, *green* to Ag, *blue* to Pt and *red* to Pd). Scale bars are 200 nm

making a Ni and a Au ring around a polymeric core (Fig. 4.1c). Then, the polymer was dissolved with acetone, and the Ni segment was etched with 3 % $FeCl_3$ to generate a gold ring within the pores at a deliberately chosen location (Fig. 4.10a). P-type P3HT was then polymerized within the pores, and the pores were subsequently widened by etching with aqueous 0.5 M NaOH. The n-type semiconductor (CdSe) was then electrochemically deposited to surround the ring-encased p-type semiconductor. At the end of this wire, a gold segment was electrodeposited as an electrical contact to generate the targeted device architecture. A z-contrast image of a typical nanowire consisting of a P3HT core (diameter: 120 nm) surrounded by a 175 nm wide, 75 nm long gold ring, further encased in CdSe (outer diameter: 235 nm) clearly shows that all components are where they were programmed to be (Fig. 4.10b).

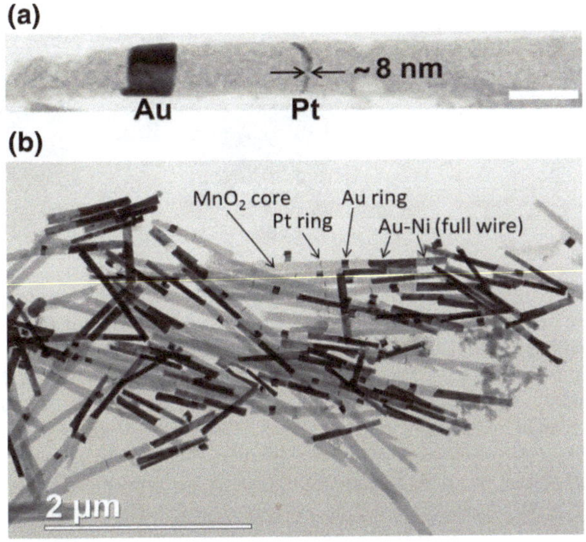

Fig. 4.9 Sub-10 nm resolution achieved by COAL. **a** TEM image of a MnO_2 core-Au ring/Pt ring nanowire with a 8 nm thick Pt ring. Scale bar: 100 nm. **b** Large scale image showing a collection of the Ni/Au/MnO_2 core-Au ring/Pt ring nanowires (ring outer diameter: 80 nm, ring inner diameter: 40 nm)

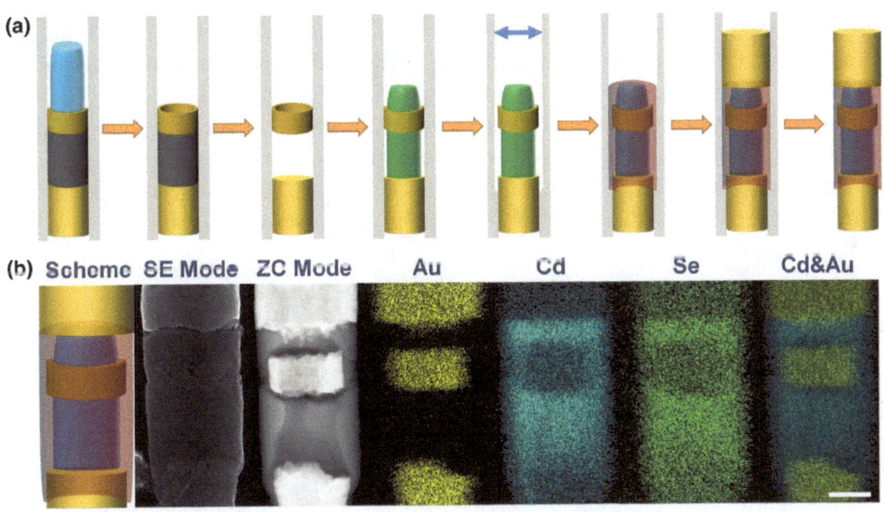

Fig. 4.10 Integration of a plasmonic gold ring within a hybrid junction composed of an organic p-type core (P3HT) and an inorganic n-type shell (CdSe). **a** Scheme illustrating the modified synthesis steps. From *left* to *right* dissolution of the PANI core (shown in *blue*), etching of the sacrificial segments (*grey*) within the AAO template, deposition of the P3HT core (*green*), pore-widening step, growth of the CdSe shell (*red*) around the P3HT core and the Au ring (*yellow*), deposition of the top Au segment and dissolution of the AAO template. **b** SE and ZC STEM images, and elemental maps of the P3HT core/CdSe shell nanowires with a Au ring (*yellow* Au, *blue* Cd, and *green* Se). Scale bar is 100 nm for all of the images

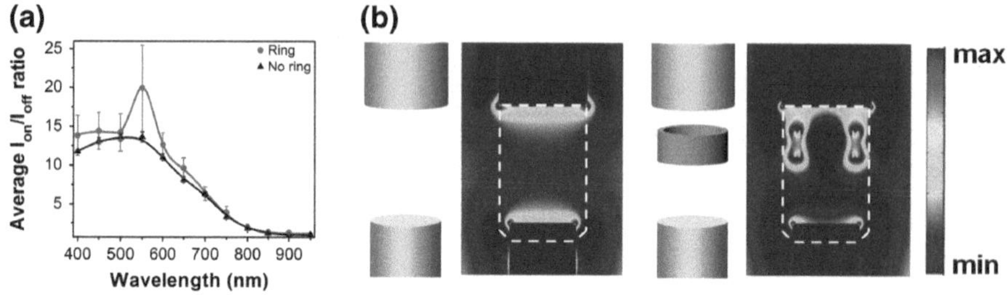

Fig. 4.11 Preliminary results with embedded rings. **a** Comparison of the average I_{on}/I_{off} ratios as a function of wavelength of the nanowires with (*red circles*) and without a ring (*black triangles*). Three nanowires were measured in each case (with and without a ring). The error bars are the standard errors of the experimental measurements. The *red* and *black lines* are only guides for the eye. **b** Simulated electric-field intensity maps of the metal segments (nanowire shown in **b**), without (*left*) and with (*right*) a ring, recorded at 532 nm (logarithmic scale). The maps were generated using an excitation source polarized in the direction parallel to the longitudinal axis of the nanowires. The dotted line corresponds to the location of the semiconductor segments (see color figure online)

The effect of the embedded plasmonic ring on the photoconductive properties of the core/shell semiconductor nanowire was investigated by preliminary electrical measurements under vacuum. Nanowires made in this manner and similar structures without a gold ring were irradiated using a xenon lamp with a monochromator while the I-V characteristics were measured (Fig. 4.11a). The plasmonic ring modifies the photoresponse of the nanowires as evidenced by the rise in current at the wavelength of the surface plasmon resonance of the gold ring. Indeed, the I_{on}/I_{off} at this wavelength was on average 45 % higher than for structures without the gold rings (3 wires averaged for each). Finite-difference time-domain (FDTD) simulations (Fig. 4.11b) suggest that this enhancement arises from the greatly increased electric field within and around the gold ring at 532 nm (for simplicity, the semiconductor segment was not included in the model, however, a red shift is expected due to the higher dielectric constants of the semiconductors).

4.3 Conclusions

In this chapter, we have reported a novel technique that bridges templated synthesis and lithography to generate coaxial nanowires in a parallel fashion with structural control in multiple dimensions and with materials generality. Using this technique, approximately 10^9 well-defined coaxial nanowires can be synthesized in parallel using commercially available AAO membranes (diameter of 2.5 cm) within a reasonable amount of time (1–16 h per synthesis). This makes COAL compatible with most laboratory-scale applications and the use of larger membranes would allow for the scaling up of this process. COAL will be extremely useful for prototyping device architectures that demand components and features not easily made by any existing technique, and as such should become a valuable research tool in the nanophotonics, energy harvesting, and nanotechnology fields for studying fundamental light-matter interactions.

4.4 Methods and Materials

4.4.1 Materials and Instruments

All chemicals and solutions were used without further processing. Commercially available plating solutions (Cyless for Ag, Orotemp 24 Rack for Au, Pallaspeed VHS for Pd, and nickel sulfamate for Ni) were purchased from Technic Inc., USA. Thiophene (≥99 %), 3-hexylthiophene (≥99 %), cadmium sulfate (99 %), lithium perchlorate (99.99 %), selenium dioxide (99.9 %), boron trifluoride diethyl etherate, cadmium chloride (99.99 %), sulfur (≥99.5 %), dimethyl sulfoxide (≥99 %), aniline (≥99.5 %), potassium hydroxide (≥99 %), concentrated perchloric acid (≥99.999 %), nitric acid (ACS grade), ammonium hexachloroplatinate (99.999 %), sodium phosphate dibasic (99 %), sulfuric acid (ACS grade) and sodium citrate (≥99 %) were purchased from Sigma Aldrich, USA. Manganese acetate was obtained from Alfa. Nanopure™ water was used. Porous anodized aluminum oxide (AAO) membranes with nominal pore diameters of 280 nm were purchased from Whatman Inc., USA. AAO membranes with 35, 55 and 100 nm nominal pore diameters were purchased from Synkera Technologies Inc., USA.

Secondary electron (SE mode) and high-angle annular dark-field imaging z-contrast (ZC mode) scanning transmission electron microscope (STEM) images were acquired using a Hitachi HD-2300 STEM. Electrochemical deposition of metals and inorganic semiconductors were done using a BASi EC epsilon potentiostat (Bioanalytical Systems, Inc., USA). Instruments and methods used for the EDX maps and for the single nanowire measurements are described later in the text.

4.4.2 Nanowire Synthesis

Porous anodized aluminum oxide (AAO) membranes with nominal pore diameters of 280 nm (Whatman Inc., USA) and of 35, 55, 100 nm (Synkera Technologies Inc., USA) were used as templates to synthesize nanowires in a three electrode setup, as described in detail elsewhere [10, 25, 26]. The bottom side of the membranes was coated with a 200 nm thick Ag layer using either thermal or e-beam evaporation. Unless mentioned otherwise, a conventional three-electrode electrochemical cell was used for the material electrodeposition, with a Pt counter electrode, an Ag/AgCl reference electrode, while the Ag-coated AAO membrane was used as the working electrode. The electrochemical cell and electrodes were thoroughly washed in nanopure water after each deposition steps.

Electrochemical depositions

Gold: Au was deposited at −930 mV (280 and 100 nm template) and −1100 mV (55 and 35 nm template) using Orotemp 24 Rack solution.

Silver: Ag was deposited at −900 mV using Cyless solution.

Palladium: Pd was deposited at −900 mV using Pallaspeed VHS solution.

Nickel: Ni was deposited at −930 mV (280 and 100 nm template) and −1100 mV (55 and 35 nm template) using nickel sulfamate solution.

Platinum: Pt was deposited at −520 mV using a homemade aqueous solution using 15 mM $(NH_4)_2PtCl_6$ and 200 mM Na_2HPO_4.

Iron: Fe was deposited at −1100 mV using a homemade Fe solution using 600 mg $FeSO_4{\cdot}7(H_2O)$, 1200 mg H_3BO_3, 20 mg $C_6H_8O_6$ dissolved in 40 mL of water.

Polypyrrole (PPy): PPy was deposited at +750 mV, using a homemade solution containing 510 μL of pyrrole dissolved in 30 mL of a 0.1 M $LiClO_4$ aqueous solution.

Polyaniline (PANI): PANI was deposited at +1000 mV, using a homemade solution containing 680 μL of aniline dissolved in a 15 mL 0.1 M $HClO_4$ aqueous solution.

Polythiophene (PTh) and poly(3-hexylthiophene) (P3HT): PTh and P3HT were deposited using cyclic voltammetry between −400 and +1100 mV at 400 mV/s. A Pt rod was used as the reference electrode. The monomers were dissolved in boron trifluoride diethyl etherate (BFEE) which served as the solvent and the electrolyte [10, 29]. Prior to the deposition, the electrochemical cell and the AAO membrane were immersed in ethanol and dried under vacuum to remove any residual water.

Cadmium Selenide: CdSe was deposited as previously reported using cyclic voltammetry between −387 and −787 mV versus SCE at 752 mV/s [31]. The plating solution was composed of 0.7 mM SeO_2, 0.3 M $CdSO_4$, and 0.25 M H_2SO_4. Triton X (0.25 % v/v) was added to the solution.

Cadmium Sulfide: CdS was deposited as previously reported under constant current (−1.5 mA/cm^2) at 130 °C in a two electrode configuration [21]. The plating solution was made by dissolving 1.52 g of $CdCl_2$ and 914 mg of S in 150 mL of hot DMSO.

Manganese Oxide: MnO_2 was deposited according to the literature at +750 mV using an aqueous solution of manganese acetate (49 mg of manganese acetate was dissolved in 20 mL of water) [32].

Approach #1: Synthesis of a polymer core with metal nanorings

Ag, Ni and then Au were successively deposited within the AAO membrane. Following the deposition of the metal segments, the polymer core was deposited. The samples were vacuum dried for 30 min to create empty spaces between the polymer segment and the wall of the AAO pore. This space was filled up with alternating layers of the metals of interest to generate a multi-segmented shell around the polymeric core. For example, the nanowires shown in Fig. 4.1c were produced by using sacrificial Ni rings, while Au was used as the target material to create Au rings around a PPy core. Following the deposition of the nanowire, the

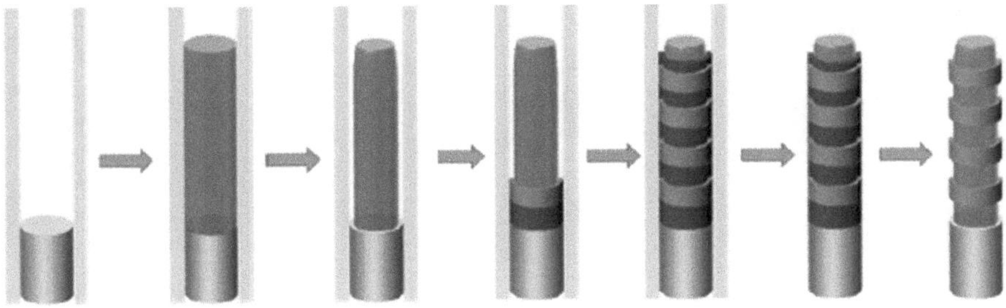

Fig. 4.12 General scheme showing the synthesis of metal nanorings around an organic core. *Blue* organic core. *Yellow* target material (Au). *Grey* sacrificial material (Ni) (see color figure online)

Ag backing layer was etched in a 4:1:1 ethanol:ammonium hydroxide:hydrogen peroxide solution for 20 min. The AAO membrane was then dissolved in 0.5 M NaOH for 10 min under continuous shaking to release the nanowires. Nanowires were spun down 4 min at 2000 for larger wires and 7500 rpm for smaller wires (soft acceleration and deceleration mode of an Eppendorf 5417R microcentrifuge). Lower spinning speeds were used for the large diameter wires to avoid unwanted breaking of the segments. The wires were then rinsed 3 times in H_2O (0.1 % sodium citrate by weight). Finally, the sacrificial segments were etched (when using a sacrificial Ni shell segments, a 50 % nitric acid aqueous solution was used for 10 min). The nanowires were then washed and spun down three times as previously described (Fig. 4.12).

Approach #2: Pore widening to control the diameter of the nanorings and/or synthesize core/shell/shell nanowires

Synthesis of the Ag-Ni-Au-polymer nanowires was performed as described in approach #1 (Fig. 4.13). Following the polymer core deposition step, the membrane was vacuum dried for 30 min. Pore widening was performed by exposing the membrane to 0.5 M NaOH. The outer diameter of the nanoring was controlled by the pore widening time (from 0 to 18 min). Similarly, pore widening can be done after the deposition of the first shell segment to fabricate a core-shell-shell nanowire, such as the nanowire shown in Fig. 4.4.

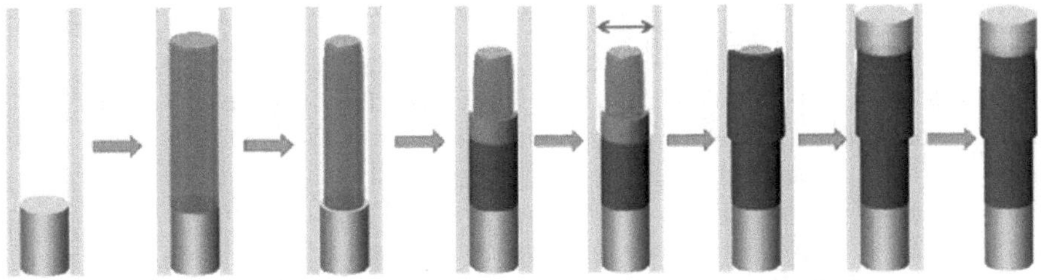

Fig. 4.13 Scheme showing the control over the segment diameter obtained via pore widening. *Blue* organic core. *Yellow* inorganic material 1 (Au). *Grey* inorganic material 2 (Ni) (see color figure online)

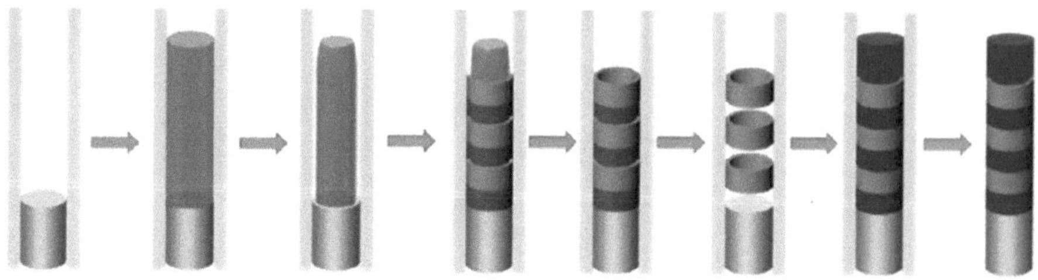

Fig. 4.14 General scheme of COAL. *Blue* sacrificial organic core (PANI). *Purple* any conductive material. *Yellow* target material (Au). *Grey* sacrificial material (Ni) (see color figure online)

Approach #3: Generalization of COAL to non-shrinking inorganic materials

Synthesis of the Ag-Ni-Au-PANI nanowires with the desired shell segments was performed as described in approach #1. Following the shell deposition, the polymer core was dissolved in a suitable solvent, sacrificial metal segments were etched in the AAO membrane, and the inorganic material was finally deposited. For example, the nanowire shown in Fig. 4.7 was composed of a PANI core (pore diameter: 55 nm) with alternating shell segments of Ni and Au. After the Au/Ni shell deposition, the PANI core was dissolved in acetone for 6 h (the acetone solution was exchanged several times). The Ni shell segments were then etched by immersing the top of the AAO membrane in a 3 % $FeCl_3$ aqueous solution for 1 h. The inorganic core (composed of MnO_2) was then deposited under constant potential. Nanowires were released into the solution using the same procedure described in approach #1 (Fig. 4.14).

Approach # 4: Integration of metal nanorings within a core-shell semiconductor nanowire

A combination of the approaches described above was used to fabricate the nanowire shown in Fig. 4.10. Following the deposition of a Ag-Ni-Au multisegmented nanorod, PANI was electropolymerized and dried under vacuum for 30 min in the pores of the AAO membrane (pore diameter: 100 nm). Alternating segments of Ni and Au were successively deposited. PANI was dissolved in acetone for 6 h and the sacrificial Ni segments were etched by immersing the AAO membrane in a 3 % $FeCl_3$ aqueous solution for 1 h. The membrane was then dried under vacuum for 30 min prior to the deposition of the P3HT core to remove any residual solvent. The P3HT was then electropolymerized, and grows through the Au ring. Pore widening was then performed to create room for the growth of the CdSe shell, while the location of the Au ring was maintained due to the very strong mechanical bond between the polymeric core and the Au ring. After the CdSe deposition, which grows around the P3HT core and the Au ring, a final Au segment was deposited on top to create the top Au electrode. Nanowires were released into solution using the same procedure described in approach #1 (Fig. 4.15).

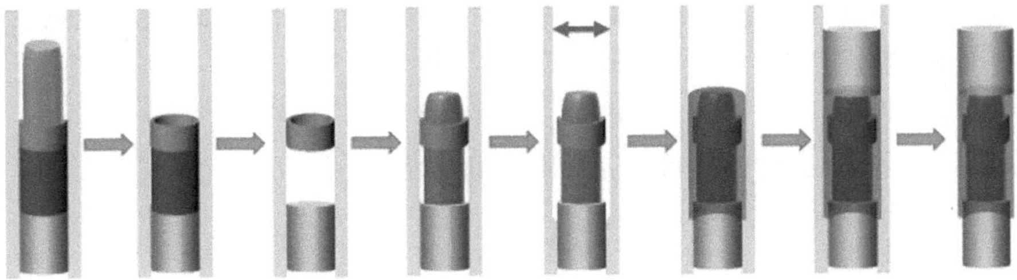

Fig. 4.15 Modified COAL process to generate metal nanorings within core-shell nanowires. *Blue* sacrificial PANI core. *Yellow* target material (Au). *Grey* sacrificial material (Ni). *Green* organic semiconductor (P3HT). *Red* inorganic semiconductor shell (CdSe) (see color figure online)

4.4.3 Electrical Characterization

Si wafer with a 500 nm oxide coating was spin-coated at 500 rpm for 10 s and at 4000 rpm for 40 s with a layer of S1805 photoresist (Shipley, USA) and was annealed at 115 °C for 1 min. Patterning on the resist was made using a Microtech MA6 Aligner mask aligner (Süss, Germany) and the patterns were developed with MF-24A (Microchem, USA) for 1 min. For the electrode pad fabrication, 5 nm of Cr and 100 nm of Au were evaporated and the photoresist and excess metal layer were lifted off using Remover PG (Microchem, USA) overnight. Multi-segmented nanowires were drop-casted on the patterned Si chips on a hot plate at 70 °C and left for drying for 5 min. Metal electrodes on the Si chip and the nanorod electrode segments on the nanowires were bridged using Quanta FESEM (FEI, USA) electron beam lithography (EBL). Si wafers with well dispersed nanowires were spin-coated at 500 rpm for 10 s and at 3000 rpm for 45 s with a layer of 950 PMMA C7 e-beam resist (Microchem, USA) and annealed at 180 °C for 2 min. Fine patterning was done using the Nanometer Pattern Generation System (NPGS, JC Nabity Lithography System, Bozeman, MT, USA) at 30 kV acceleration voltage and the patterns were developed with 3:1 IPA/MIBK solution for 1 min. 3 nm of Cr and 75 nm of Au films were evaporated and the excess materials were lifted off overnight in acetone.

The electrical measurements were carried out under vacuum ($\sim 10^{-5}$ Torr) using a Keithley 4200-SCS semiconductor characterization system and a probe station. Current-voltage measurements on single nanowires were performed under dark and under white light using the built-in microscope lamp as the illumination source. A 300 W xenon arc lamp (Newport) was passed through an Oriel 1/8 m 77,250 monochromator, and the monochromatic output light was carried to the sample with a fiber optic cable to serve as the excitation source for the spectral photocurrent measurements. The output power was measured using a S130C slim photodiode power sensor connected to PM200 power and energy meter console from Thorlabs at a collection wavelength matching the value set on the monochromator. Nanowires were exposed to monochromatic light for 10 s over the 400–950 nm range with a 50 nm step size. The top three values recorded during light exposure were averaged and divided by the value of the electrical current in the absence of illumination to calculate the light on/light off ratio for each measurement. The I_{on}/I_{off} ratios were plotted as a function of wavelength.

4.4.4 Finite-Difference Time-Domain Simulations

Electric field intensity generated by the plasmonic metal segments (with and without the ring) were calculated and plotted using a commercially available finite-difference time-domain (FDTD) simulation software package developed by Lumerical Solutions Inc., Vancouver, Canada. Simulations were done with the use of a total-field scattered-field plane wave source polarized in the direction parallel to the longitudinal axis of the nanowires to detect the effect of enhanced electric fields directed into the semiconducting region. The refractive index of the medium was set to 1 (experimental measurements were carried under vacuum). Electric field calculations were done in 3D (x-y-z), and 1 nm mesh size was used for the calculations. Johnson and Christy data set was used directly from the Lumerical Software materials library for gold.

4.4.5 Elemental Mapping

Energy Dispersive X-Ray Spectroscopy (EDS) mapping was performed using a Hitachi HD-2300 STEM equipped with two EDS Oxford detectors.

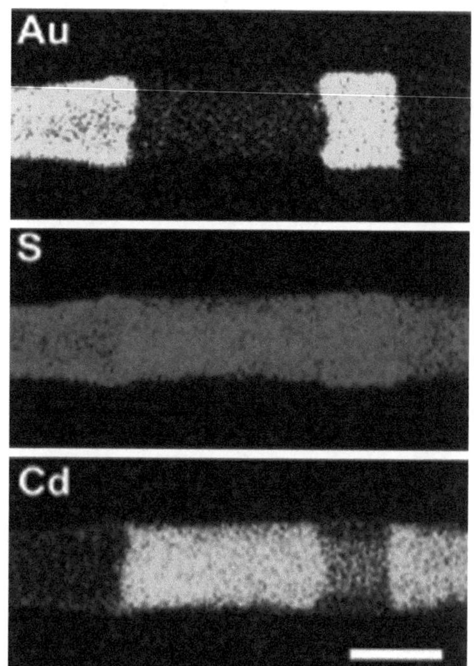

Fig. 4.16 EDS maps of the CdS core-Au ring nanowire shown in Fig. 4.8b. *Yellow* Au $L_{\alpha 1}$ line at 9.712 keV (integration: 9.726–10.042 keV), *purple* sulfur K_{α} line at 2.307 keV (integration: 2.193–2.421 keV), and *blue* Cd $L_{\alpha 1}$ line at 3.133 keV (integration: 3.010–3.256 keV). Sulfur is detected where the gold and the CdS segments are located because of the overlap between the Au $M_{\beta,\gamma}$ lines (2.204 and 2.410 keV, respectively, not used for the mapping of Au) and the sulfur K_{α} line at 2.307 keV (see color figure online)

For the MnO_2/Au nanowire presented in Fig. 4.8a, we used the Au $M_{\alpha 1}$ line at 2.122 keV (integration: 2.014–2.230 keV) and the Mn K_{α} line at 5.895 keV (integration: 5.746–6.044 keV).

For the CdS/Au nanowire presented in Fig. 4.8b, we used the Au $L_{\alpha 1}$ line at 9.712 keV (integration: 9.726–10.042 keV) and the Cd $L_{\alpha 1}$ line at 3.133 keV (integration: 3.010–3.256 keV).

For the Ni/Au/Ag/Pt/Pd nanowire presented in Fig. 4.8c, we used the Ni $L_{\alpha 1}$ line at 0.851 keV (integration: 0.754–0.948 keV), Au $M_{\alpha 1}$ line at 2.122 keV (integration: 2.014–2.230 keV), Ag $L_{\alpha 1}$ line at 2.984 keV (integration: 2.862–3.106 keV), Pt $L_{\alpha 1}$ line at 9.441 keV and Pt $L_{\alpha 2}$ line at 9.360 keV (integration: 9.251–9.528 keV) and Pd $L_{\alpha 1}$ line at 2.838 keV (integration: 2.718–2.958 keV).

As can be seen on this map (Fig. 4.8c), there is some overlap between the Au and Pt signals due to some overlap between the Au $L_{\alpha 2}$ line (9.626 keV) and the Au $M_{\alpha 1}$ line (2.122 keV) with the Pt $L_{\alpha 1}$ line (9.441 keV) and the Pt $M_{\alpha 1}$ line (2.050 keV), respectively.

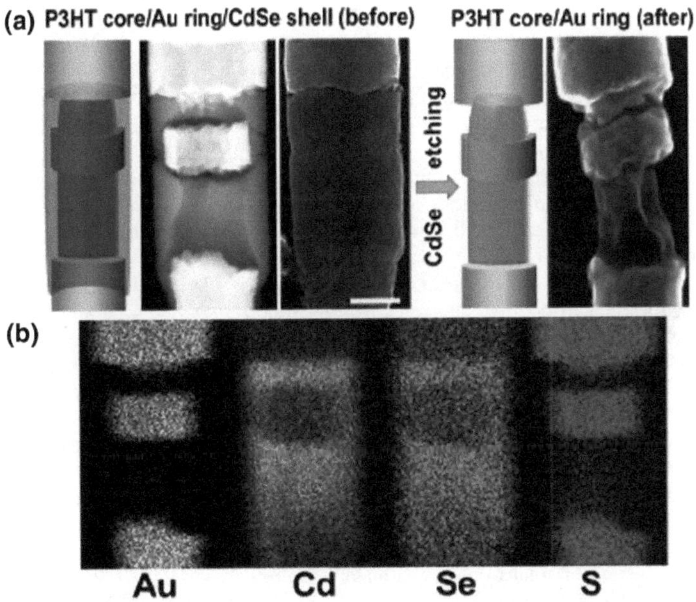

Fig. 4.17 EDS maps of the P3HT core-Au ring-CdSe shell nanowire shown in Fig. 4.10b. **a** (*Left*) Scheme, ZC and SE mode STEM images of the P3HT core-Au ring-CdSe shell nanowire shown in Fig. 4.10b. (*Right*) Scheme and typical SE mode STEM image of the nanowires after etching of the CdSe shell with concentrated nitric acid, revealing the polymeric P3HT core. **b** EDS maps of the nanowire before etching the CdSe shell. *Yellow* Au $L_{\alpha 1}$ line at 9.712 keV (integration: 9.726–10.042 keV), *blue* Cd $L_{\alpha 1}$ line at 3.133 keV (integration: 3.010–3.256 keV), *green* Se $L_{\alpha 1}$ line at 1.379 keV (integration: 1.275–1483 keV) and *red* sulfur K_{α} line at 2.307 keV (integration: 2.193–2.421 keV). It is clear that the S signal is located on the P3HT core and on the Au segments because of the overlap between the Au $M_{\beta,\gamma}$ lines (2.204 and 2.410 keV, respectively, not used for the mapping of Au) and the sulfur K_{α} line at 2.307 keV (see color figure online)

For similar reasons, there is some overlap between the Ag and the Pd signals due to the some overlap between the Ag $L_{\alpha 1}$ line at 2.984 keV and the Pd $L_{\alpha 1}$ line at 2.838 keV. We verified that the rings were made of pure element by doing point EDS measurements (not shown here).

For the P3HT core-Au ring-CdSe shell nanowire shown in Fig. 4.10b, we used the Au $L_{\alpha 1}$ line at 9.712 keV (integration: 9.726–10.042 keV), Cd $L_{\alpha 1}$ line at 3.133 keV (integration: 3.010–3.256 keV), and Se $L_{\alpha 1}$ line at 1.379 keV (integration: 1.275–1483 keV).

For clarity and due to the overlap between the Au $M_{\beta,\gamma}$ lines (2.204 and 2.410 keV, respectively) and the S K_{α} line (2.307 keV), we did not include the sulfur maps originating from the CdS core (Fig. 4.8b) and the P3HT core (Fig. 4.10b) in the main text. However, we have included both of them in Figs. 4.16 and 4.17.

References

1. Garnett, E. C., Brongersma, M. L., Cui, Y., & McGehee, M. D. (2011). Nanowire solar cells. *Annual Review of Materials Research, 41*, 269.
2. Jones, M. R., Osberg, K. D., Macfarlane, R. J., Langille, M. R., & Mirkin, C. A. (2011). Templated techniques for the synthesis and assembly of plasmonic nanostructures. *Chemical Reviews, 111*, 3736.
3. Stewart, M. E., Anderton, C. R., Thompson, L. B., Maria, J., Gray, S. K., Rogers, J. A., et al. (2008). Nanostructured plasmonic sensors. *Chemical Reviews, 108*, 494.
4. Ozbay, E. (2006). Plasmonics: merging photonics and electronics at nanoscale dimensions. *Science, 311*, 189.
5. Tang, J., Huo, Z., Brittman, S., Gao, H., & Yang, P. (2011). Solution-processed core-shell nanowires for efficient photovoltaic cells. *Nature Nanotechnology, 6*, 568.
6. Aricò, A. S., Bruce, P., Scrosati, B., Tarascon, J.-M., & Van Schalkwijk, W. (2005). Nanostructured materials for advanced energy conversion and storage devices. *Nature Materials, 4*, 366.
7. Wallentin, J., Anttu, N., Asoli, D., Huffman, M., Åberg, I., Magnusson, M. H., et al. (2013). InP nanowire array solar cells achieving 13.8% efficiency by exceeding the ray optics limit. *Science, 339*, 1057.
8. Atwater, H. A., & Polman, A. (2010). Plasmonics for improved photovoltaic devices. *Nature Materials, 9*(3), 205–213.
9. Linic, S., Christopher, P., & Ingram, D. B. (2011). Plasmonic-metal nanostructures for efficient conversion of solar to chemical energy. *Nature Materials, 10*, 911.
10. Bourret, G. R., Ozel, T., Blaber, M., Shade, C. M., Schatz, G. C., & Mirkin, C. A. (2013). Long-range plasmophore rulers. *Nano Letters, 13*, 2270.
11. Brittman, S., Gao, H., Garnett, E. C., & Yang, P. (2011). Absorption of light in a single-nanowire silicon solar cell decorated with an octahedral silver nanocrystal. *Nano Letters, 11*, 5189.
12. Piner, R. D., Zhu, J., Xu, F., Hong, S., & Mirkin, C. A. (1999). "Dip-Pen" nanolithography. *Science, 283*, 661.
13. Qin, L., Park, S., Huang, L., & Mirkin, C. A. (2005). On-wire lithography. *Science, 309*, 113.
14. Chou, S. Y., Krauss, P. R., & Renstrom, P. J. (1995). Imprint of sub-25 nm vias and trenches in polymers. *Applied Physics Letters, 67*, 3114.

15. Mubeen, S., Lee, J., Singh, N., Krämer, S., Stucky, G. D., & Moskovits, M. (2013). An autonomous photosynthetic device in which all charge carriers derive from surface plasmons. *Nature Nanotechnology, 8*, 247.
16. Tian, B., Zheng, X., Kempa, T. J., Fang, Y., Yu, N., Yu, G., et al. (2007). Coaxial silicon nanowires as solar cells and nanoelectronic power sources. *Nature, 449*, 885.
17. Dillen, D. C., Kim, K., Liu, E.-S., & Tutuc, E. (2014). Radial modulation doping in core-shell nanowires. *Nature Nanotechnology, 9*, 116.
18. Zhu, F., Chern, G., Tchernyshyov, O., Zhu, X., Zhu, J., & Chien, C. (2006). Magnetic bistability and controllable reversal of asymmetric ferromagnetic nanorings. *Physical Review Letters, 96*, 027205.
19. Penner, R. M., & Martin, C. R. (1987). Preparation and electrochemical characterization of ultramicroelectrode ensembles. *Analytical Chemistry, 59*, 2625.
20. Martin, C. R. (1994). Nanomaterials: a membrane-based synthetic approach. *Science, 266*, 1961.
21. Routkevitch, D., Bigioni, T., Moskovits, M., & Xu, J. M. (1996). Electrochemical fabrication of CdS nanowire arrays in porous anodic aluminum oxide templates. *Journal of Physical Chemistry, 100*, 14037.
22. Martin, B. R., Dermody, D. J., Reiss, B. D., Fang, M. M., Lyon, L. A., Natan, M. J., et al. (1999). Orthogonal self-assembly on colloidal gold-platinum nanorods. *Advanced Materials, 11*, 1021.
23. Nicewarner-Pena, S. R., Freeman, R. G., Reiss, B. D., He, L., Pena, D. J., Walton, I. D., et al. (2001). Submicrometer metallic barcodes. *Science, 294*, 137.
24. Mirkovic, T., Foo, M. L., Arsenault, A. C., Fournier-Bidoz, S., Zacharia, N. S., & Ozin, G. A. (2007). Hinged nanorods made using a chemical approach to flexible nanostructures. *Nature Nanotechnology, 2*, 565.
25. Osberg, K. D., Schmucker, A. L., Senesi, A. J., & Mirkin, C. A. (2011). One-dimensional nanorod arrays: independent control of composition, length, and interparticle spacing with nanometer precision. *Nano Letters, 11*, 820.
26. Banholzer, M. J., Qin, L., Millstone, J. E., Osberg, K. D., & Mirkin, C. A. (2009). On-wire lithography: synthesis, encoding and biological applications. *Nature Protocols, 4*, 838.
27. Qin, L., Jang, J. W., Huang, L., & Mirkin, C. A. (2007). Sub-5-nm gaps prepared by on-wire lithography: correlating gap size with electrical transport. *Small, 3*, 86.
28. Gao, H., Liu, C., Jeong, H. E., & Yang, P. (2011). Plasmon-enhanced photocatalytic activity of iron oxide on gold nanopillars. *ACS Nano, 6*, 234.
29. Ozel, T., Bourret, G. R., Schmucker, A. L., Brown, K. A., & Mirkin, C. A. (2013). Hybrid semiconductor core-shell nanowires with tunable plasmonic nanoantennas. *Advanced Materials, 25*, 4515.
30. Chen, X., Li, S., Xue, C., Banholzer, M. J., Schatz, G. C., & Mirkin, C. A. (2008). Plasmonic focusing in rod−sheath heteronanostructures. *ACS Nano, 3*, 87.
31. Schierhorn, M., Boettcher, S. W., Kraemer, S., Stucky, G. D., & Moskovits, M. (2009). Photoelectrochemical performance of CdSe nanorod arrays grown on a transparent conducting substrate. *Nano Letters, 9*, 3262.
32. Liu, R., & Lee, S. B. (2008). MnO2/poly (3, 4-ethylenedioxythiophene) coaxial nanowires by one-step coelectrodeposition for electrochemical energy storage. *Journal of the American Chemical Society, 130*, 2942.

Chapter 5
Solution Dispersible Metal Nanorings: Independent Control of Architectural Parameters and Materials Generality

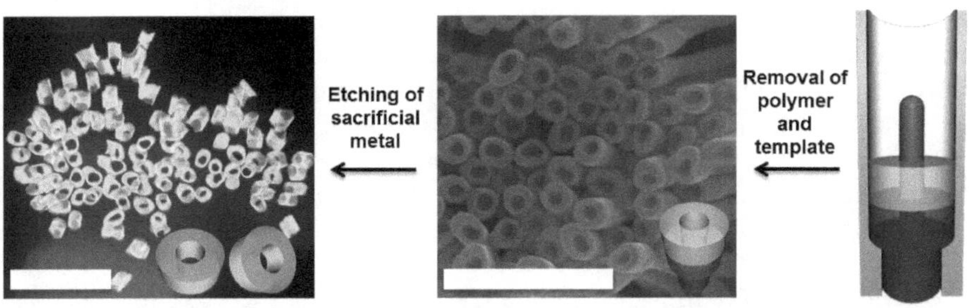

5.1 Introduction

Advances in lithographic and synthetic techniques have enabled the fabrication of a variety of complex structures on the nanoscale [1, 2]. In particular, the synthesis of metal nanoparticles has remained a major field of research in the last few decades since such particles show extraordinary optical properties compared to their bulk counterparts [1–4]. Small alterations in their size, shape, composition, and the surrounding dielectric medium change these properties, resulting in exceptional spectral tunability [4]. Such spectral engineering of nanoparticles is strongly tied to the oscillation of conduction band electrons within the metal nanoparticle; this phenomenon is described by an emerging field in photonics research known as plasmonics. Researchers have focused on studying plasmonic interactions in a number of metal-metal [1, 3, 5–7], metal-semiconductor [8–13], and metal-insulator [3, 4, 14, 15] platforms, and have shown their uses in a range of applications from

---

Portions of this chapter have been published in ***Nano Letters*** 15 (8), 5273 (2015).

Co-authors of this work: Michael. J. Ashley, Michael B. Ross, Gilles R. Bourret, George C. Schatz, Chad A. Mirkin.

T. Ozel, *Coaxial Lithography*, Springer Theses,
DOI 10.1007/978-3-319-45414-6_5

surface-enhanced spectroscopies [14] to nanoscale rulers [9] and from plasmonic metamaterials [16] to light focusers [6, 17]. Recent improvements in synthetic and measurement techniques have led to remarkable progress in the understanding of how metal nanostructures interact with light. One optically interesting but synthetically challenging architecture is the nanoring, whose study requires precise control over three different architectural parameters (i.e. inner diameter, outer diameter, and length) which all affect the three different plasmonic modes observed. Previous studies in light-nanoring interactions have been conducted using metal nanorings that were patterned on transparent substrates using colloidal lithography [18–22], $AgNO_3$ heat treatment [23], nanoimprint lithography [24, 25], and electron beam lithography [26, 27]. Lately, however, there has been a great interest in preparing metal nanorings on larger scales and in solution so that all of the plasmonic resonances in the metal nanorings could be excited with a single unpolarized light source and ensemble-averaged measurements could be reported. Moreover, solution dispersible nanorings have been favored for their ease of processing for uses in different fundamental and device application studies. Previously employed processes for solution dispersible nanoring fabrication have included galvanic replacement of Co [28] or Ag [29] (as an extension of On-Wire Lithography [7]), the use of a molecular [30] or anodized [31] template, polyol synthesis [32, 33], and other chemical methods [34–36]. However, there are still a limited number of synthetic methods that are capable of controlling all of the architectural parameters in a rational way. These methods also lack compatibility with a variety of metals and a wide range of dimensions (from a few tens of nanometers to a few microns). To the best of our knowledge, there is no synthetic method, including electron beam lithography and wet chemistry, which allows precise control of all of the three architectural parameters that could be used to prepare rings with a variety of single composition metals or bimetallic rings separated in either axial or radial dimensions. In a recent publication, we presented how COAL can be used to synthesize core-shell nanowires with combinations of several metals and semiconductors with sub-10 nm resolution in both the axial and radial dimensions.

5.2 Results and Discussion

In this work, we present the use of COAL methodology to prepare metal nanorings with high precision in all three architectural parameters (with diameters ranging from 25 to 400 nm), as well as the compatibility of this technique with a variety of different single composition metals (including Au, Ag, Pt). Furthermore, we systematically varied each of the architectural parameters of Au nanorings and studied the resulting changes in their spectral properties through ensemble-averaged extinction measurements and discrete-dipole approximation calculations.

Nanoring synthesis was performed within porous anodized aluminum oxide (AAO) templates by sequential electrodeposition of target and sacrificial materials in a conventional three-electrode electrochemical setup [5, 7, 13, 37]. A 350 nm silver film was initially evaporated onto one side of the AAO template, which was then placed into an electrochemical cell for a short pore-refinement step by exposing the template to 0.5 M NaOH solution for up to 3 min (depending on the diameter of the template, see Supporting Information for details). We have found that in addition to increasing the average diameter of the AAO pores (allowing us access to all pore diameters intermediate to stock templates), this step serves to increase the uniformity of pore shape and size. Electrodeposition began with a silver nanorod segment to ensure a good conductive interface between the evaporated silver film and the subsequent metal segments. A sacrificial nickel segment was then electrodeposited, functioning as a favorable base for the subsequent electropolymerization of aniline, which occurs under constant potential to prepare polyaniline (PANI) nanorods inside the template. PANI nanorods were then dried under vacuum for prompt contraction, creating space between the polymer rods and the pore walls of the template [8, 13]. The electrodeposition steps discussed so far were performed to prepare a dynamic template for the size-controllable deposition of metal nanorings in which the outer diameter of the nanorings can be defined by the diameter of the pore size of the AAO template and the inner diameter of the nanorings can be defined by the diameter of the polymer nanorod (Fig. 5.1, top row). Within the template, the outer diameter of the nanorings was controlled by widening the pores (through slow dissolution of the template from inside out using 0.5 M NaOH) under various exposure times between 3 and 18 min depending on the desired outer diameter. Alternatively, the inner diameter of the nanorings was controlled by thinning the polymer nanorods (through slow dissolution of the polymer from all sides using an ethanol/water mixture) under various exposure times between 10 and 30 min depending on the desired inner diameter. Extended periods of exposure (or exposure to overly concentrated ethanol solutions) resulted in nearly complete dissolution of the polymer nanorods, yielding a mixture of rings and disks. Following the outer and inner diameter adjustment steps, a short sacrificial metal shell segment was deposited before the target ring material in order to minimize variations in the diameter of the nanorings. A target shell segment (the desired material for the nanoring) was electrodeposited to complete the synthesis of nanorings (sequential steps of sacrificial and target shell deposition can be repeated to multiply the number of nanorings synthesized per template). Finally, the length of the nanorings was controlled by increasing the charge limit for the electrodeposition (by probing the amount of charge passed through the reference electrode) under constant potential. The nanorings can be dispersed into solution (Fig. 5.1, bottom row) by dissolving the polymer in acetone, removing the AAO template in 1.5 M NaOH, and the sacrificial nickel segment in aqueous 3 % $FeCl_3$.

The synthetic process described in Fig. 5.1 was used to prepare solution dispersible gold nanorings in templates with different nominal pore sizes (35, 55, 100 nm from Synkera Technologies). Following the electrochemical synthesis of gold nanorings, top view SEM images were taken of rings on top of the sacrificial Ni and Ag segments, with PANI and the AAO template having been dissolved (Fig. 5.2a). Measurements performed on the top view SEM images (50+ nanorings were measured for each sample) show that nanoring synthesis in the templates

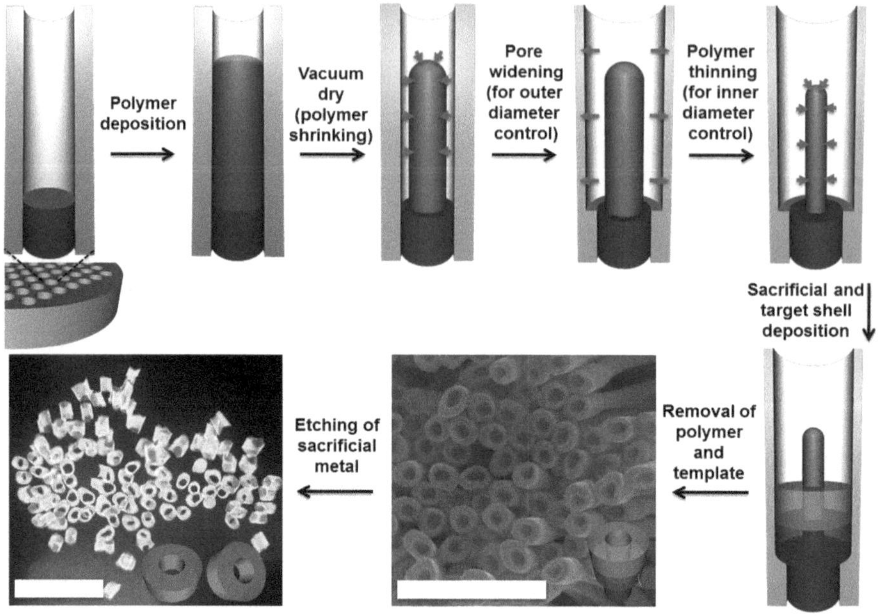

Fig. 5.1 Nanoring synthesis. Fabrication scheme (*top row* and *bottom right*, represented by the cross-section of a single AAO pore) shows the synthetic steps involving electrodeposition of a polymer nanorod on a metal segment followed by polymer contraction under vacuum drying to create space (~5–10 nm) between it and the pore walls. Both the final outer and inner diameters of the nanorings were determined prior to the deposition of shell segments. A pore-widening step was performed to increase the diameter of the pores (corresponding to the outer diameter of the nanorings) whereas a polymer thinning step was performed to decrease the diameter of the polymer nanorods (corresponding to the inner diameter of the nanorings). Following the deposition of sacrificial and target shell segments, the polymer segments and the template were removed and nanorings were imaged while still attached to sacrificial metal segments (*top view* scanning electron microscopy image, *bottom middle*). Nanorings were then released into solution by etching these sacrificial metal segments (scanning transmission microscopy image, *bottom left*). Scale bars equal to 1 μm

mentioned above yields to gold nanorings with diameters of 63 ± 5, 95 ± 10, and 156 ± 16 nm. These nanorings were then released into solution by removing the sacrificial metal segments for solution-averaged extinction measurements to determine their plasmonic properties (Fig. 5.2b). The flexibility of using different templates with different diameters (course tuning of the architectural parameters) resulted in a wide-range tailorable surface plasmon resonance peaks from visible to near-IR range (593–1110 nm only with the change of the template diameter).

Fine tuning of the architectural parameters are summarized in Fig. 5.3 by showing the extent to which nanoring size (and consequent surface plasmon resonance spectrum) can be tuned specifically within the 100 nm nominal diameter templates. Our base case sample was pore-widened in 0.5 M NaOH for 18 min, underwent no ethanol treatment, and short in length (aspect ratio of 0.4). This was

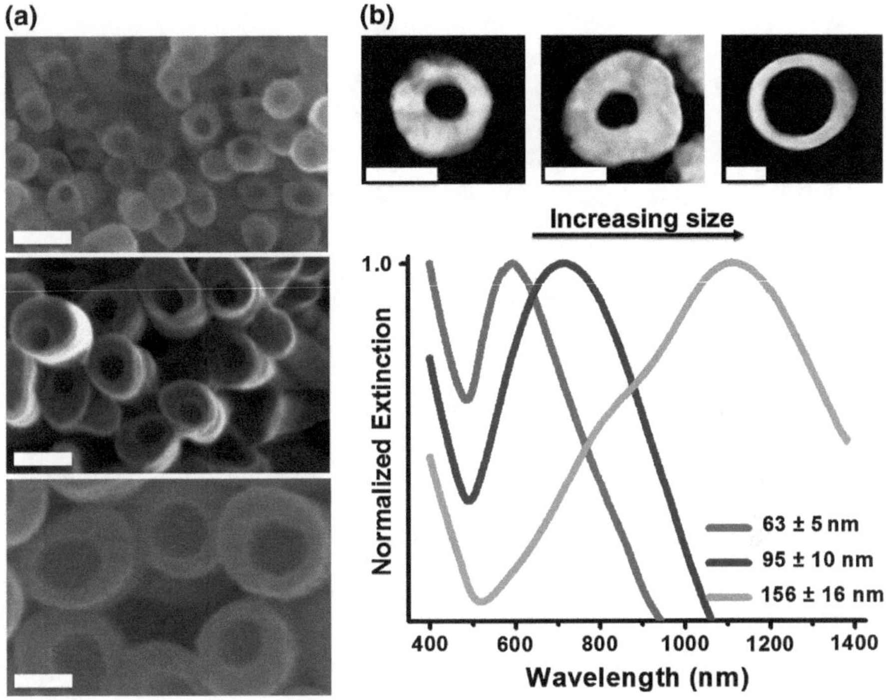

Fig. 5.2 Gold nanorings synthesized in templates with different diameters. **a** *Top view* electron microscope image of gold nanorings with diameters ~63, 95, 156 nm (*top* to *bottom*, respectively). Scale bars equal to 100 nm. **b** In-solution extinction spectra of gold nanorings with different diameters (shown in **a** and *top row*) show plasmon resonance peaks at ~593, 715, 1110 nm (smaller to largest template, respectively). Scale bars equal to 50 nm

determined to be the maximum pore-widening time, corresponding to the maximum achievable nanoring outer diameter with this template. NaOH exposure beyond 18 min led to the merging of some pores, compromising the homogeneity of the resulting solution. Conversely, pore-widening must be performed for at least 3 min in order to guarantee a solution of fully-formed rings without the presence of some crescent-like structures. Synthesizing solutions at each of these extremes allowed for the demonstration of outer diameter tenability using AAO templates with equivalent nominal diameters. Figure 5.3a shows that an average outer diameter increase from 156 to 179 nm, corresponding to a slight red-shift in the main plasmon resonance wavelength by 42 nm (from 1113 to 1155 nm), can be realized through pore-widening, with inner diameter and aspect ratio remaining nearly constant.

The second deviation from the base case sought to investigate the extent and effects of inner diameter control. This was achieved by the slow dissolution of the PANI core that defines the inner region of the resulting nanorings. However, before this took place, samples were pore-widened for 18 min in order to maximize space

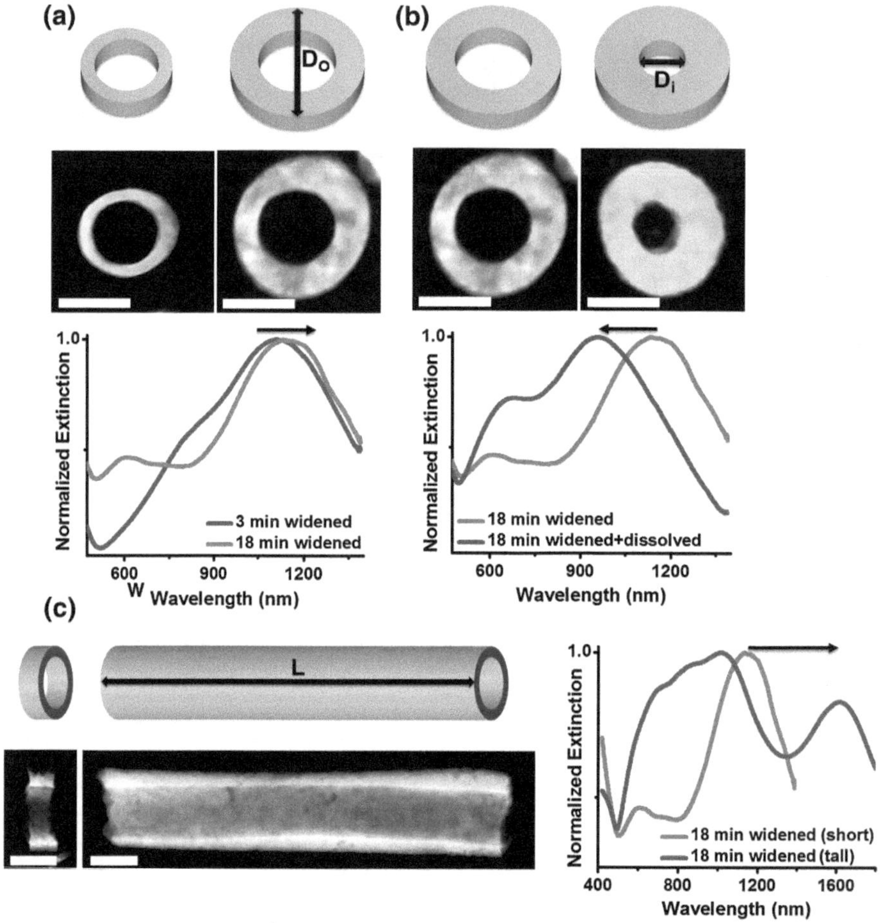

Fig. 5.3 Control in all three architectural parameters of a gold nanoring. **a** The outer diameter enlarges by increasing the duration of the pore-widening step from 3 to 18 min. This increases the outer diameter by 25 nm, which corresponds to a slight *red-shift* in the main plasmon resonance wavelength by 42 nm. **b** The inner diameter decreases with an increasing duration of the polymer thinning step. By the introduction of the polymer thinning step, the inner diameter was reduced by 31 nm, which corresponds to a significant *blue-shift* in the plasmon resonance wavelength by 197 nm. **c** The length of the nanoring increases with the amount of charge passing through the electrodes during the electrodeposition step. By increasing the applied charge from 350 to 2500 mC, aspect ratio was increased from ~0.4 to ~5, which corresponds to a significant *redshift* in the lowest frequency plasmon resonance wavelength by 462 nm. Scale bars equal to 100 nm in all of the electron microscopy images. *Blue curve* in all of the extinction spectra correspond to the same nanoring which have been pore-widened for 18 min (see color figure online)

for the ethanol/water solution to travel deep into the pores, increasing PANI dissolution from its sides. After a series of PANI dissolution experiments testing the potency of different concentrations of ethanol in water (60–100 %) for varied periods of time (10–110 min), it was found that the rate of polymer thinning slows with exposure time. Accordingly, it was determined that using aqueous solutions with ethanol concentrations above 70 % or for over 30 min was unnecessary, as these thinned the polymer nearly as much as stronger solutions, but did not dissolve it as much from the top. Thus, we observed the smallest nanoring inner diameters (while precluding the introduction of disks into solution, caused by excessive PANI dissolution from the top) by exposing samples to 70 % ethanol for 30 min. Figure 5.3b shows that average inner diameter can be reduced from 95 to 64 nm (while maintaining consistent outer diameter and aspect ratio), corresponding to a significant blue-shift in the plasmon resonance wavelength by 197 nm (from 1155 to 958 nm), by using this polymer thinning technique.

The final architectural parameter to be programmed was nanoring length. Control over the length of a material is well-established in electrodeposition [5, 7, 37–42], one simply adjusts the amount of charge passed through the electrodes and tunes to the desired length. However, this investigation represents a novel look into how the length of a shell material can be varied and offers insight into the viability of tube synthesis using COAL. The aspect ratios (length to outer diameter) of the samples discussed so far were deliberately kept around 0.4 (except for the smallest nanoring sample, whose aspect ratio was 0.75 so that structural stability could be guaranteed) in order to eliminate the convoluting effects that an inconsistent aspect ratio would have on the resulting optical properties of nanorings. That way, any surface plasmon resonance (SPR) wavelength shifts could be solely attributed to the varied architectural parameter. In this case, the nanoring aspect ratio was adjusted to be $\sim$5 by substantially increasing the charge limit of the system, resulting a significant redshift in the lowest frequency plasmon resonance wavelength by 462 nm (from 1155 to 1617 nm, Fig. 5.3c). The outer and inner diameters were consistent with the base case sample, such that length was the only parameter changed.

One of the main advantages of using COAL for the synthesis of metal nanorings is its materials generality. The synthetic approach presented above is a universal approach for the synthesis of nanorings composed of any electrodepositable material (as long as the target nanoring material is stable under exposure to the sacrificial segment remover). To demonstrate this point, we synthesized metal nanorings made of Au, Pt, and Ag and characterized them using EDX (Fig. 5.4). Au and Pt rings were synthesized by depositing a Ni sacrificial segment followed by either Au or Pt. However, for the protection of Ag rings during the sacrificial Ag film removal (evaporated on one side of the template as the working electrode), Ag rings were synthesized by depositing alternating Ni-Ag-Ni shell segments. Sacrificial Ni segments were then removed using dilute $FeCl_3$ or phosphoric acid solutions. Elemental analysis measured from the red spot on the nanoring images confirmed the composition of the nanorings.

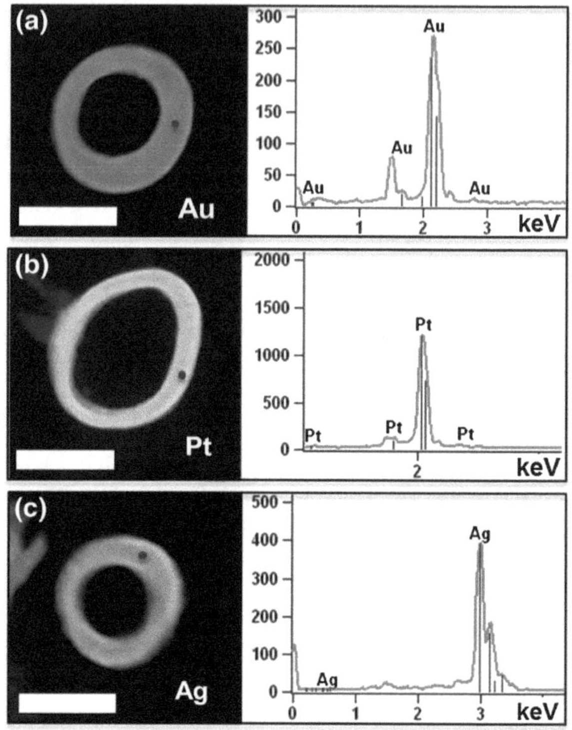

Fig. 5.4 Synthesis of nanorings with different materials. STEM images (*false colored*, *left*) and elemental analysis (*right*) measured at the *red* spot located on the STEM images of **a** gold, **b** platinum, and **c** silver nanorings (see color figure online)

5.3 Conclusions

Dynamic template preparation via COAL is a unique tool for the synthesis of nanorings with an unprecedented control in architectural and compositional parameters. This technique allows for the synthesis of nanorings in a wide range of templates with different diameters and offers additional control over all of the intermediate architectural parameters such as the outer diameter, inner diameter, and length, for precise tailoring of the nanoring dimensions. This fine architectural tuning results in exceptional control over the optical properties of the nanorings. Also, the synthetic procedure presented in this work is a universal approach for templated nanoring synthesis and can be performed with any electrochemically depositable material (that is chemically stable during the removal of the sacrificial segments).

5.4 Methods and Materials

5.4.1 Materials and Instruments

All chemicals and solutions were used without further processing. Commercially available plating solutions (Cyless for Ag, Orotemp 24 Rack for Au, Pallaspeed VHS for Pd, and nickel sulfamate for Ni) were purchased from Technic Inc., USA. Aniline (≥99.5 %), concentrated perchloric acid (99.999 %), ammonium hexachloroplatinate (99.999 %), sodium phosphate dibasic (99 %), NH_4OH (28–30 %), and sodium citrate (≥99 %) were purchased from Sigma Aldrich, USA. 30 % aqueous H_2O_2 (30 %) was purchased from VWR, USA. AAO membranes with 35, 55 and 100 nm nominal pore diameters were purchased from Synkera Technologies Inc., USA.

Secondary electron (SE mode) and high-angle annular dark-field imaging z-contrast (ZC mode) scanning transmission electron microscope (STEM) images were acquired using a Hitachi HD-2300 STEM. Electrochemical deposition of metals and inorganic semiconductors were done using a BASi EC epsilon potentiostat (Bioanalytical Systems, Inc., USA).

<u>Electrochemical Depositions</u>
Metals were deposited at constant potentials using aqueous plating solutions.

Au was deposited at −950 mV (280 and 100 nm template) and −1100 mV (55 and 35 nm template) using Orotemp 24 Rack solution.

Ag was deposited at −910 mV using Cyless solution.

Pd was deposited at −900 mV using Pallaspeed VHS solution.

Ni was deposited at −930 mV (280 and 100 nm template) and −1100 mV (55 and 35 nm template) using nickel sulfamate solution.

Pt was deposited at −520 mV using a homemade aqueous Pt solution (15 mM $(NH_4)_2PtCl_6$ and 200 mM Na_2HPO_4).

Fe was deposited at −1000 mV using a homemade aqueous Fe solution (5 mg ascorbic acid, 150 mg iron sulfate heptahydrate, and 300 mg boric acid in 10 ml H_2O).

Polyaniline (PANI): PANI was deposited at +1000 mV, using a homemade solution containing 680 μL of aniline dissolved in a 15 mL 0.1 M $HClO_4$ aqueous solution.

<u>Nanoring Characterization</u>
Outer and inner diameters of the Au rings were measured using Adobe Photoshop CS5. Top view SEM images were taken of rings on top of the sacrificial Ni and Ag segments, with PANI and the AAO template having been dissolved. Traces were drawn around the outer and inner surfaces of each ring (sample size = 50), yielding measurements of the area and perimeter of both in nm^2 and nm, respectively. From these measurements, both the outer and inner diameters of rings could be calculated; this eliminated the bias of choosing where to measure these lengths (for imperfect rings) as would exist if measuring in a linear fashion.

5.4.2 Nanoring Synthesis

Templated nanoring synthesis began with porous anodized aluminum oxide (AAO) membranes coated with a 350 nm thick Ag layer. For the 35 and 55 nm template syntheses, AAO is first exposed to 0.5 M NaOH for 2 min. This step (termed "pore refinement") serves to modestly enlarge the pore size of the smaller AAO templates to guarantee that the subsequent steps can be reproducibly performed. Electrochemical deposition was carried out using a conventional three-electrode setup, as described elsewhere in detail [5, 37]. First, Ag and Ni were successively deposited within the AAO membrane. Next, a polyaniline (PANI) core was deposited and samples were dried under light vacuum for 30 min to create space between the polymer segments and the walls of the AAO pores. As the Ni base was more conductive than the PANI core, subsequent electrodeposition led to this space being filled from the bottom up with our materials of interest. However, before shell deposition, both the final outer and inner diameters of the nanorings were determined. A pore-widening step was first conducted in which the AAO membrane was exposed to 0.5 M NaOH for 3–18 min. In this step, the NaOH solution traveled down into the AAO pores and slowly dissolved the template from the inside out. This created more space between PANI and the pore walls, ensuring that fully-formed ring structures were later synthesized (a minimum of 3 min was required for this to be reliable), and allowing for outer diameter tunability. Pore-widening for beyond 18 min, however, resulted in partial merging of some pores and inconsistent shell deposition. Particularly for 35 nm AAO templates, the initial pore refinement step was critical to ensure that every subsequent step to this point occurred with complete reproducibility. Without it, PANI was not guaranteed to contract enough to allow sufficient space for NaOH to widen the pores such that full rings were consistently generated. We have also observed that pore refinement improves the uniformity of AAO pores. In order to achieve inner diameter control, the template was first pore-widened (18 min for the 100 nm template sample with inner diameter reduction, and 10 min for the 35 nm template sample; inner diameter reduction was not performed on the 55 nm template sample) and then exposed to a 70 % ethanol solution in water for 30 min. Within each pore, ethanol slowly dissolved PANI from all sides such that the core material diminished in size and the resulting rings had a reduced inner diameter. The previous pore-widening step was necessary as it increased the amount of ethanol that PANI was exposed to on its sides relative to its top, maximizing inner diameter reduction. The use of a 70 % ethanol solution for 30 min was found to result in the smallest average inner diameter for the rings without the resulting solution being composed of a mixture of rings and disks. When PANI was exposed to ethanol either in a more concentrated solution or for longer times, the (several micron long) polymer segment dissolved too much from the top, such that the following electrodeposited material reached beyond it in some of the pores and formed a disk. Additionally, ethanol solutions of varied concentration for varied time periods can be used to achieve nanorings with inner diameters intermediate to those reported in the main text. After pore-widening and (optional) polymer thinning with ethanol, sequential deposition of a sacrificial shell segment (typically Ni) and a target material shell segment (the desired nanoring) finalize the templated portion of the synthesis. The amount of charge passed during electrodeposition controlled shell segment length, and so final ring length. Au and Pt rings were synthesized by depositing a Ni sacrificial segment followed by Au/Pt. Ag rings were synthesized by depositing alternating Ni-Ag-Ni shell segments. Pd rings were synthesized by depositing an Fe-Pd-Fe-Ni shell (Fe serves to prevent alloying between Pd and Ni and Ni protects Pd and Fe from the subsequent Ag etchant). In all cases, the Ag backing layer was then dissolved using a mixture of 30 % aqueous H_2O_2 (Macron, Inc.) and 28 % aqueous NH_4OH (Sigma-Aldrich) (1:1 v:v). After immersing in water for 5 min, AAO membranes were placed in acetone for at least 3 h in order to fully dissolve PANI. Templates were then dissolved in 1.5 M NaOH (with 0.5 mM sodium citrate and 0.05 mM BSPP for Ag and Pd samples). The remaining structures were then centrifuged at 9000 rpm for 3 min, and sacrificial materials were removed with 3 % $FeCl_3$ in water for 15 min for Au and Pt rings, 15 % H_3PO_4 in water for 5 min for Ag rings, and 50 % H_3PO_4 for 20 min for Pd rings. Nanorings were centrifuged at 10,500 rpm for 6 min 3 times and resuspended in water (0.1 % sodium citrate by weight for Au and Pt, 0.5 mM sodium citrate and 0.05 mM BSPP for Ag and Pd samples).

References

1. Jones, M. R., Osberg, K. D., Macfarlane, R. J., Langille, M. R., & Mirkin, C. A. (2011). Templated techniques for the synthesis and assembly of plasmonic nanostructures. *Chemical Reviews, 111*, 3736.
2. Stewart, M. E., Anderton, C. R., Thompson, L. B., Maria, J., Gray, S. K., Rogers, J. A., et al. (2008). Nanostructured plasmonic sensors. *Chemical Reviews, 108*, 494.
3. Ozbay, E. (2006). Plasmonics: merging photonics and electronics at nanoscale dimensions. *Science, 311*, 189.
4. Kelly, K. L., Coronado, E., Zhao, L. L., & Schatz, G. C. (2003). The optical properties of metal nanoparticles: the influence of size, shape, and dielectric environment. *The Journal of Physical Chemistry B, 107*, 668.
5. Osberg, K. D., Schmucker, A. L., Senesi, A. J., & Mirkin, C. A. (2011). One-dimensional nanorod arrays: independent control of composition, length, and interparticle spacing with nanometer precision. *Nano Letters, 11*, 820.
6. Nordlander, P. (2009). The ring: a leitmotif in plasmonics. *ACS Nano, 3*, 488.
7. Qin, L., Park, S., Huang, L., & Mirkin, C. A. (2005). On-wire Lithography. *Science, 309*, 113.
8. Ozel, T., Bourret, G. R., Schmucker, A. L., Brown, K. A., & Mirkin, C. A. (2013). Hybrid semiconductor core-shell nanowires with tunable plasmonic nanoantennas. *Advanced Materials, 25*, 4515.
9. Bourret, G. R., Ozel, T., Blaber, M., Shade, C. M., Schatz, G. C., & Mirkin, C. A. (2013). Long-range plasmophore rulers. *Nano Letters, 13*, 2270.
10. Ozel, T., Nizamoglu, S., Sefunc, M. A., Samarskaya, O., Ozel, I. O., Mutlugun, E., et al. (2011). Anisotropic emission from multilayered plasmon resonator nanocomposites of isotropic semiconductor quantum dots. *ACS Nano, 5*, 1328.

11. Brittman, S., Gao, H., Garnett, E. C., & Yang, P. (2011). Absorption of light in a single-nanowire silicon solar cell decorated with an octahedral silver nanocrystal. *Nano Letters, 11*, 5189.
12. Atwater, H. A., & Polman, A. (2010). Plasmonics for improved photovoltaic devices. *Nature Materials, 9*, 205.
13. Ozel, T., Bourret, G. R., & Mirkin, C. A. (2015). Coaxial lithography. *Nature Nanotechnology, 10*, 319.
14. Willets, K. A., & Van Duyne, R. P. (2007). Localized surface plasmon resonance spectroscopy and sensing. *Annual Review of Physical Chemistry, 58*, 267.
15. Jensen, T. R., Duval, M. L., Kelly, K. L., Lazarides, A. A., Schatz, G. C., & Van Duyne, R. P. (1999). Nanosphere lithography: effect of the external dielectric medium on the surface plasmon resonance spectrum of a periodic array of silver nanoparticles. *The Journal of Physical Chemistry B, 103*, 9846.
16. Henzie, J., Lee, M. H., & Odom, T. W. (2007). Multiscale patterning of plasmonic metamaterials. *Nature Nanotechnology, 2*, 549.
17. Babayan, Y., McMahon, J. M., Li, S., Gray, S. K., Schatz, G. C., & Odom, T. W. (2009). Confining Standing Waves in Optical Corrals. *ACS Nano, 3*, 615.
18. Aizpurua, J., Hanarp, P., Sutherland, D., Käll, M., Bryant, G., & García de Abajo, F. (2003). Optical properties of gold nanorings. *Physical Review Letters, 90*, 057401.
19. McLellan, J. M., Geissler, M., & Xia, Y. (2004). Edge spreading lithography and its application to the fabrication of mesoscopic gold and silver rings. *Journal of the American Chemical Society, 126*, 10830.
20. Larsson, E. M., Alegret, J., Käll, M., & Sutherland, D. S. (2007). Sensing characteristics of NIR localized surface plasmon resonances in gold nanorings for application as ultrasensitive biosensors. *Nano Letters, 7*, 1256.
21. Sun, Z., Li, Y., Zhang, J., Li, Y., Zhao, Z., Zhang, K., et al. (2008). A Universal Approach to Fabricate Various Nanoring Arrays Based on a Colloidal-Crystal-Assisted-Lithography Strategy. *Advanced Functional Materials, 18*, 4036.
22. Yan, F., & Goedel, W. A. (2004). Preparation of mesoscopic gold rings using particle imprinted templates. *Nano Letters, 4*, 1193.
23. Yuan, Z.-H., Zhou, W., Duan, Y.-Q., & Bie, L.-J. (2008). A simple approach for large-area fabrication of Ag nanorings. *Nanotechnology, 19*, 075608.
24. Lee, S. H., Choi, S. M., Yoon, S., Jeong, H., Jung, G. Y., Cho, B. K., et al. (2014). Transfer printing of metal nanoring and nanodot arrays for use in catalytic reactions. *Chemical Communications*.
25. Kim, S., Jung, J.-M., Choi, D.-G., Jung, H.-T., & Yang, S.-M. (2006). Patterned arrays of Au rings for localized surface plasmon resonance. *Langmuir, 22*, 7109.
26. Banaee, M. G., & Crozier, K. B. (2010). Gold nanorings as substrates for surface-enhanced Raman scattering. *Optics Letters, 35*, 760.
27. Near, R., Tabor, C., Duan, J., Pachter, R., & El-Sayed, M. (2012). Pronounced effects of anisotropy on plasmonic properties of nanorings fabricated by electron beam lithography. *Nano Letters, 12*, 2158.
28. Hu, Y., Chou, T., Wang, H., & Du, H. (2014). Monodisperse colloidal gold nanorings: synthesis and utility for surface-enhanced raman scattering. *The Journal of Physical Chemistry C, 118*, 16011.
29. Liusman, C., Li, S., Chen, X., Wei, W., Zhang, H., Schatz, G. C., et al. (2010). Free-standing bimetallic nanorings and nanoring arrays made by on-wire lithography. *ACS Nano, 4*, 7676.
30. Zinchenko, A. A., Yoshikawa, K., & Baigl, D. (2005). DNA-templated silver nanorings. *Advanced Materials, 17*, 2820.
31. Bridges, C. R., DiCarmine, P. M., & Seferos, D. S. (2012). Gold nanotubes as sensitive, solution-suspendable refractive index reporters. *Chemistry of Materials, 24*, 963.
32. Gong, H. M., Zhou, L., Su, X. R., Xiao, S., Liu, S. D., & Wang, Q. Q. (2009). Illuminating dark plasmons of silver nanoantenna rings to enhance exciton–plasmon interactions. *Advanced Functional Materials, 19*, 298.

33. Zhou, L., Fu, X.-F., Yu, L., Zhang, X., Yu, X.-F., & Hao, Z.-H. (2009). Crystal structure and optical properties of silver nanorings. *Applied Physics Letters, 94*.
34. Kong, X. Y., Ding, Y., Yang, R., & Wang, Z. L. (2004). Single-crystal nanorings formed by epitaxial self-coiling of polar nanobelts. *Science, 303*, 1348.
35. Drogat, N., Granet, R., Sol, V., & Krausz, P. (2010). One-pot silver nanoring synthesis. *Nanoscale Research Letters, 5*, 566.
36. Jang, H.-J., Ham, S., Acapulco, J. A. I., Song, Y., Hong, S., Shuford, K. L., et al. (2014). Fabrication of 2D Au nanorings with Pt framework. *Journal of the American Chemical Society*.
37. Banholzer, M. J., Qin, L., Millstone, J. E., Osberg, K. D., & Mirkin, C. A. (2009). On-wire lithography: synthesis, encoding and biological applications. *Nature Protocols, 4*, 838.
38. Mirkovic, T., Foo, M. L., Arsenault, A. C., Fournier-Bidoz, S., Zacharia, N. S., & Ozin, G. A. (2007). Hinged nanorods made using a chemical approach to flexible nanostructures. *Nature Nanotechnology, 2*, 565.
39. Nicewarner-Pena, S. R., Freeman, R. G., Reiss, B. D., He, L., Pena, D. J., Walton, I. D., et al. (2001). *Science, 294*, 137.
40. Routkevitch, D., Bigioni, T., Moskovits, M., & Xu, J. M. (1996). Electrochemical fabrication of CdS nanowire arrays in porous anodic aluminum oxide templates. *Journal of Physical Chemistry, 100*, 14037.
41. Martin, C. R. (1994). Nanomaterials: a membrane-based synthetic approach. *Science, 266*, 1961.
42. Penner, R. M., & Martin, C. R. (1987). Preparation and electrochemical characterization of ultramicroelectrode ensembles. *Analytical Chemistry, 59*, 2625.

Chapter 6
Conclusions and Outlook on Templated Electrochemical Synthesis Using Coaxial Lithography

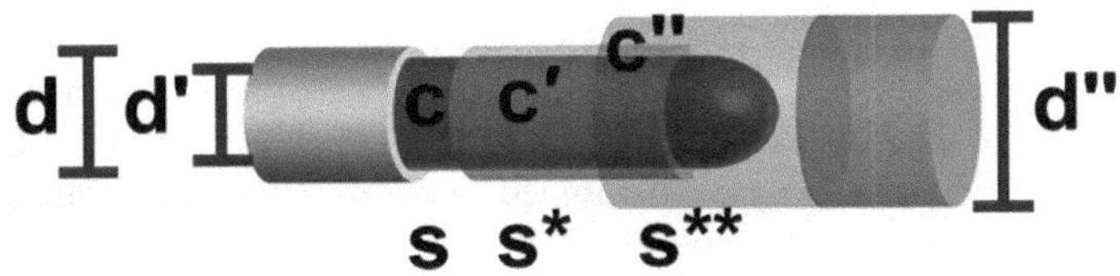

6.1 Conclusions

Existing templated nanowire syntheses and lithography methods (including On-Wire Lithography) can be used to prepare nanowires, but only with precise control over the axial dimension (i.e., 1D-control) [1–4]. Control over shell thickness and composition are necessary since they are fundamental parameters in core-shell nanowires influencing chemical stability, light absorption, and exciton recombination. With the introduction of COAL, we overcome this limitation and go well beyond current synthetic capabilities:

(i) We can control the diameter of the nanowire shell segments using a pore-widening method described in Chaps. 4 and 5, not disclosed prior to these works. Indeed, we can synthesize nanowires with sub-10 nm control over the axial and radial dimensions of the wire architectures (i.e., 2D-control). Additionally, COAL is compatible with a wide variety of materials, as shown in Chap. 4 (polymers: PANI, PPy, PTh, P3HT; metals: Au, Ag, Ni, Pt, Pd; inorganic semiconductors: MnO_2, CdSe, CdS).

(ii) Addition or deletion of materials within the radial dimension, introduced by COAL, offers the ability to engineer nanostructures along multiple directions. For example, selective etching of sacrificial shell segments allows the

Special thanks to Gilles R. Bourret, Taegon Oh, and Michael J. Ashley

T. Ozel, *Coaxial Lithography*, Springer Theses,
DOI 10.1007/978-3-319-45414-6_6

synthesis of optically and electrically decoupled metal rings with very well-defined dimensions and spacing around semiconductor nanowires. Indeed, the use of sacrificial nickel segments, which are good electrical conductors and non-reactive towards gold deposition, allows the generation of striped shells with smooth interfaces. Using this approach, an almost unlimited number of shells can be prepared with nanometer precision on the same nanowire. Complete tailorability in terms of surface plasmon resonance and electric field is reported by controlling the diameter, length, and ring spacing. This is unachievable by any other means. These advances are essential for both the nanowire and nanophotonics communities.

(iii) Finally, integration of plasmonic rings that do not block the electron-hole flow within core-shell nanowires is a one-of-a-kind design that addresses both the electrical (core-shell interface for efficient charge separation/collection) and optical properties (plasmon-enhanced absorption) in a controllable fashion (Chap. 4). The use of selective etching within the AAO membrane in conjunction with pore-widening is necessary to generate these nanowires, which could not be prepared using any other the strategy prior to COAL. Control over the radial dimension, combined with the possibility of selectively deleting features used to build the nanowires significantly expands the range of architectures that can be synthesized using COAL.

To further emphasize the importance of the architecture control possibilities afforded by COAL, I will provide a few relevant unpublished examples, which go beyond the scope of my thesis and are meant to give an outlook and show how enabling this technique will be. Indeed, I believe the COAL technique will dramatically increase the capabilities of many researchers, who are interested in studying light-matter interactions, nanoparticle assembly, solution-dispersible nanoparticles and labels, semiconductor device applications, and nanowire biomimetic probe preparation.

6.2 Light-Matter Interactions

6.2.1 Synthesis of Plasmonic Nanostructures Within and Around Semiconductor Nanowires

Compared to their bulk counterpart, noble metal nanoparticles (NPs) exhibit exceptional optical properties [5]. One important nanoscale effect is the generation of large enhanced electric fields at the metal nanoparticle surface [6]. This effect has been used to enhance the emission and absorption of semiconductors [7–9]. Until now, the only way to study the effect of plasmonic NPs on the optoelectronic properties of semiconductor nanowires was to randomly disperse NPs (usually via drop casting) onto the nanowires, and hope that the NPs would randomly end up at the right location [10, 11].

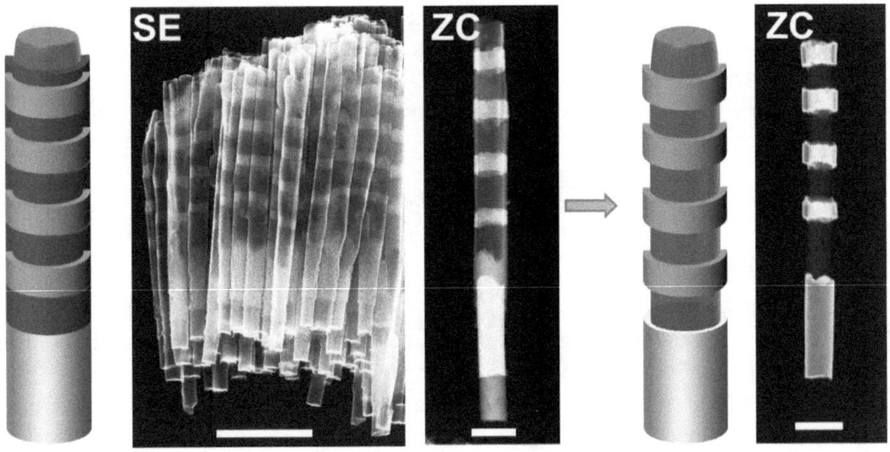

Fig. 6.1 Nanowires for plasmon-exciton interaction studies (adapted from Chap. 4). STEM images show nanowires before (*left*) and after (*right*) etching the sacrificial Ni shell to generate Au rings around a polypyrrole core. Scale bars are 2 μm, 500 nm and 500 nm, respectively

One of the strengths of COAL is its ability to directly integrate plasmonic structures with very well-defined surface plasmon resonances within semiconductor nanowires. Such a structure is presented in Fig. 6.1 (adapted from Chap. 4). The semiconductor segment (polypyrrole) is surrounded by four plasmonic nanorings (made of gold). Figure 6.2 (preliminary results) shows that we can generate

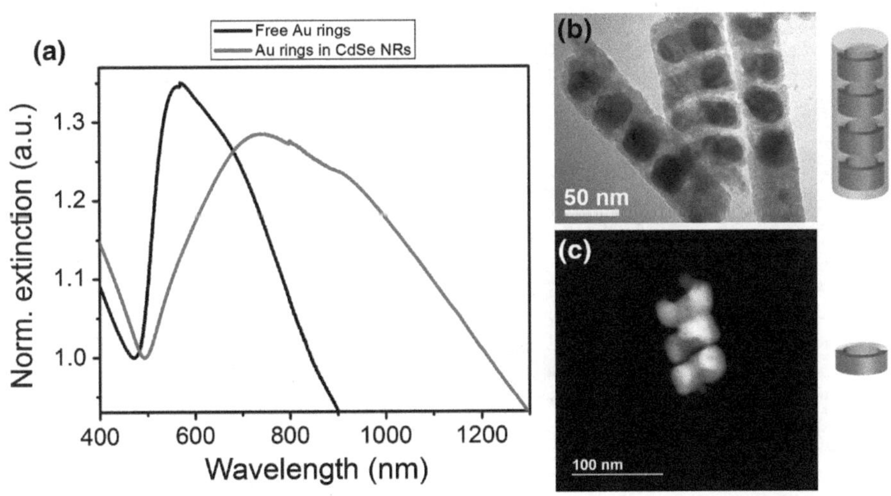

Fig. 6.2 Plasmonic rings within semiconductor nanowires (preliminary results). **a** Solution-phase extinction spectra of Au nanorings within CdSe nanorods (*red curve*), and of the free Au nanorings (*black curve*), generated after CdSe etching. **b** TEM image of the Au rings/CdSe nanorods. **c** Z-contrast STEM image of the free Au rings, generated after CdSe etching. The close proximity of the rings is due to a drying artifact (color figure online)

plasmonic rings around semiconductor nanowires with well-defined resonances from visible to near-IR. This demonstrates our ability to generate homogenous structures in terms of size and geometry. As discussed later in this chapter, these structures are excellent candidates for plasmon-exciton interaction studies.

6.2.2 *Use of COAL Enabled Structures in Photocurrent Mapping Studies*

Photocurrent mapping provides an effective way to study the manner in which plasmonic NPs can modify photocurrent generation within semiconductors. Thus far, research efforts have been focused on randomly distributed plasmonic NPs on Si nanowires. It has been shown that photocurrent can be locally enhanced at the Au NP/Si nanowire interface, around the surface plasmon resonance wavelength [10, 11]. With COAL, one can synthesize plasmonic nanorings with tunable optical resonances at specific locations around and inside semiconductor nanowires (Fig. 6.2). The structural control afforded by COAL should be highly beneficial to the preparation of plasmonic NPs/semiconductor nanowire hybrids for photocurrent generation experiments (Fig. 6.3). In particular, the ability to generate nanometer scale gaps between plasmonic rings within semiconductor nanowires could allow for new photocurrent mapping studies and improve our current understanding of plasmon-enhanced photocurrent generation.

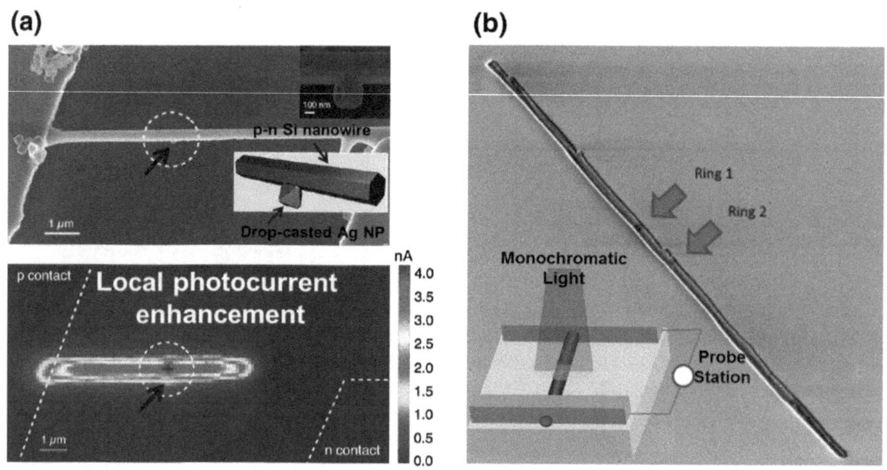

Fig. 6.3 Nanowires for plasmon-enhanced photocurrent generation studies. **a** Scheme and SEM image of a p-n Si nanowires with a Ag nanoparticle (*top*), photocurrent map of the corresponding nanowire that shows the local photocurrent enhancement around the nanoparticle (*bottom*) (adapted from Yang et al. [10]). **b** Optical microscope image of Au rings around an Fe_2O_3 core prepared using COAL, which can be used in photocurrent generation studies under monochromatic light excitation, as shown in the inset (preliminary results)

6.2.3 Integration of Plasmonic Structures Within Core-Shell Semiconductor Nanowires that Do not Block the Electron-Hole Flow

Plasmonic structures are very promising entities for improving the collection of light within photovoltaic devices [12, 13]. However, the current inability to control the position of plasmonic nanostructures within a device can lead to a local decrease in the photocurrent generation due to parasitic blocking of the electron flow within the system [13]. COAL can be used to rationally integrate tunable plasmonic rings within semiconductor heterojunctions without blocking the electron-hole flow (Fig. 6.4), by generating a structure that has not previously been made.

To show the power of COAL, we presented the integration of a plasmonic Au ring within a hybrid junction composed of an organic p-type core (P3HT) and an inorganic n-type shell (CdSe) (Fig. 6.5, adapted from Chap. 4). As suggested by the design, the electron-hole flow is not blocked and the enhanced electric fields generated by the metal nanoring results in a wavelength-dependent enhancement of the photocurrent generation, as shown and discussed in Chap. 4. By the optimization of synthetic parameters, one can synthesize multiple nanorings with different plasmon resonance frequencies to plasmonically enhance the photocurrent generation throughout the whole solar spectrum.

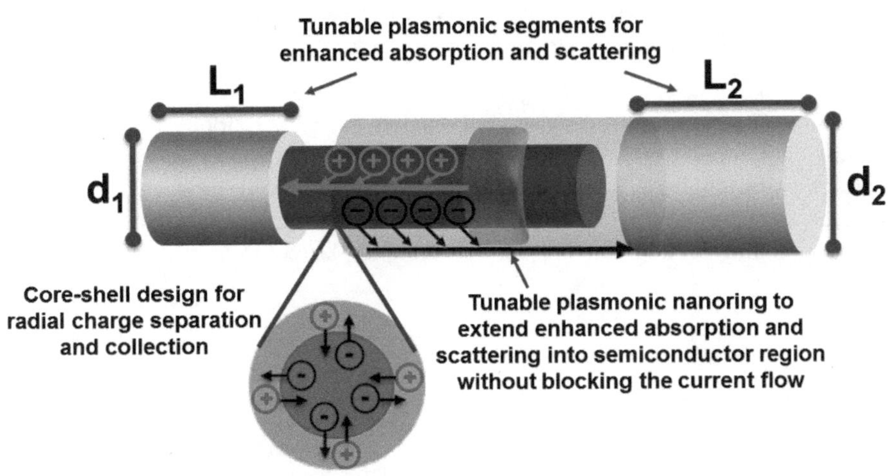

Fig. 6.4 A promising design for a light harvesting nanowire which can address the influence of both the electrical and optical properties of the nanowire on photocurrent generation

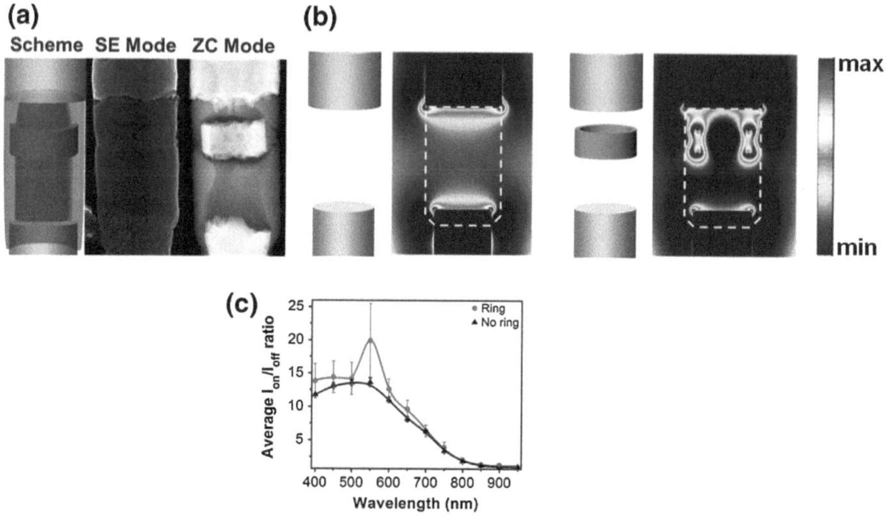

Fig. 6.5 Nanowires for plasmon-enhanced photocurrent generation (adapted from Chap. 4). **a** STEM image of a Au ring within a hybrid junction composed of an organic p-type core (P3HT) and an inorganic n-type shell (CdSe), **b** electric field intensity maps showing the enhanced fields within the semiconductor region in the presence of the Au ring, **c** wavelength dependent light on/light off current values with and without the ring

6.2.4 Synthesis of Complex Metal Nanostructures for Plasmonics Studies

More innovative architectures can be prepared via COAL for novel light-matter interaction studies. Some examples are: plasmonic gold rotaxanes ("nanogears"), SERS architectures based on the integration of a plasmonic ring cavity within a nanorod dimer to generate a very intense electric field, and concentric nanoring cavities around semiconductors for lasing and plasmonic-focusing applications (Fig. 6.6).

6.3 Nanoparticle Assembly

Nanorings can be synthesized in AAO templates using COAL as described in Chap. 5. Each ring has a sacrificial polymer segment on its inner face, sacrificial nickel segments on its top and bottom faces, and an AAO template on its outer face. These sacrificial segments can serve as nanomasks and can be removed one by one for selective functionalization of different faces of a nanoring. Depending on the functionalized face, integration of particles with complementary DNA sequence can result in assembly of these nanorings into different forms as shown in Fig. 6.7.

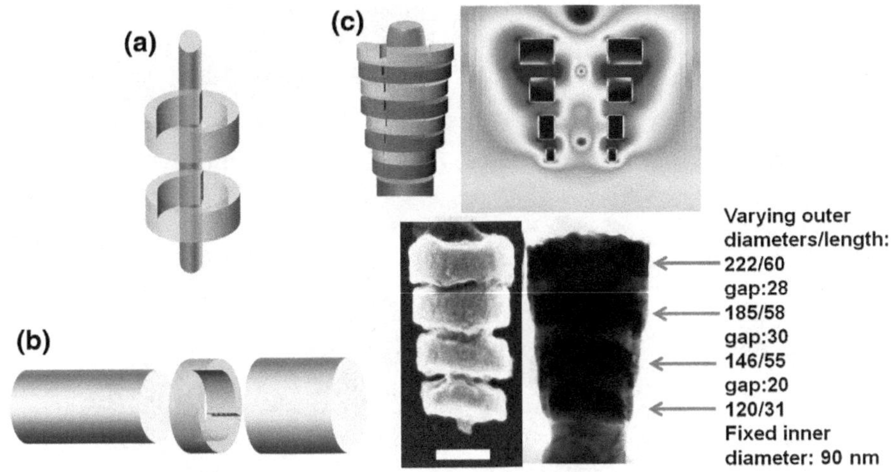

Fig. 6.6 Novel architectures enabled via COAL. Schemes of **a** nanorings around a nanobar with a gap in-between, **b** Gold nanorod dimers with a plasmonic ring in the middle for SERS, **c** concentric rings of different diameters for light focusing. **c** Electric field intensity map of four concentric gold rings with different diameters around a polymer core (*top*), and corresponding nanowire STEM image with and without the Ni segments (*bottom*), scale bar: 100 nm

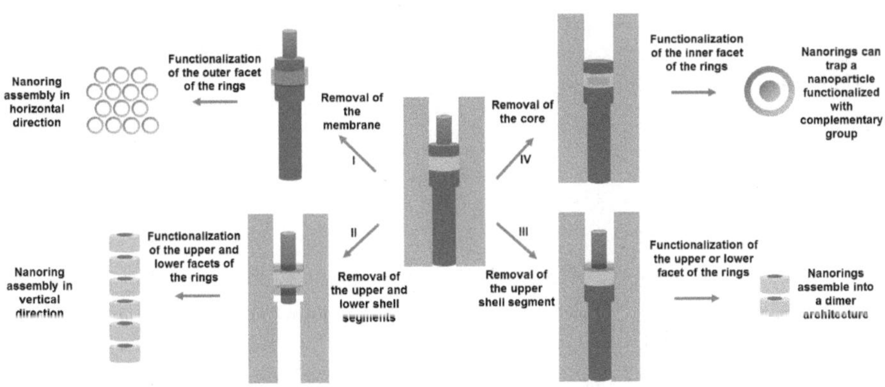

Fig. 6.7 Possible routes for the controlled assembly of metal nanostructures via COAL. (*I*) Functionalization (shown in *light red color*) of the outer face of the ring with DNA for 2D assembly. (*II*) Functionalization of the top and bottom faces of the rings using DNA for 1D assembly. (*III*) Selective functionalization of the top face of the nanoring to generate dimers. (*IV*) Selective functionalization of the nanoring inner core to assemble nanoparticles inside the ring cavity for plasmon-exciton and plasmonics studies

Preliminary experimental results (Fig. 6.8) confirm the localization of a DNA functionalized spherical gold nanoparticle inside a gold nanoring (functionalized with complementary DNA). This localized sample was prepared using COAL by following an approach similar to that shown in approach IV (Fig. 6.7). A similar

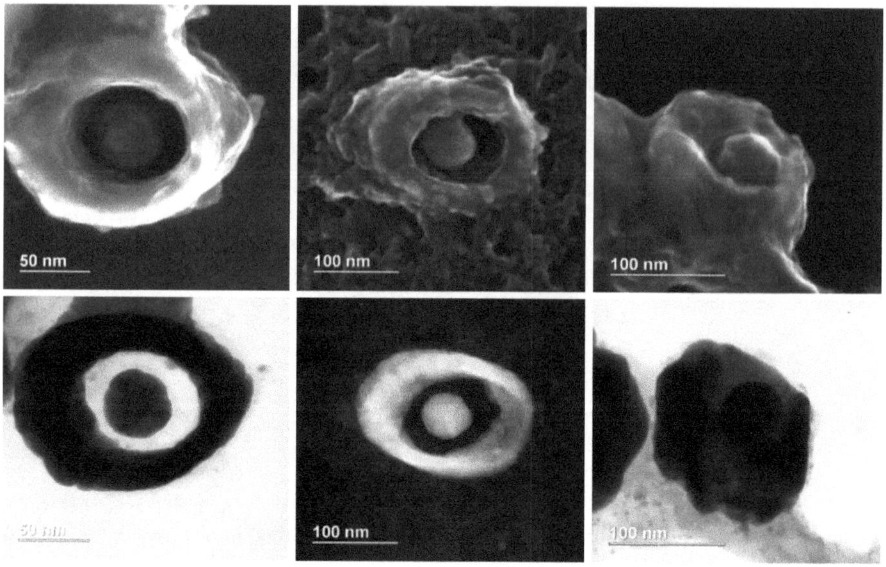

Fig. 6.8 Assembly of nanoparticles (preliminary results). STEM images of Au nanospheres trapped inside Au nanorings through complementary DNA functionalization

approach can be used for localizing a quantum dot within a plasmonic nanoring cavity. These architectures could be very useful for plasmonics, plasmon-exciton interactions, and biosensing studies.

6.4 Semiconductor Device Applications

COAL allows the synthesis of core-shell semiconductors, which we have shown to work as photodetectors [14]. By selectively etching a sacrificial nickel segment, we observed a substantial change in the electrical properties of a semiconductor nanowire (from Schottky type to p-n type current-voltage curve) (Fig. 6.9). This approach could be used to compare the photovoltaic performance of a Schottky diode and a p-n diode using the same nanowire (the effect of the sacrificial etching step on the device performance needs to be taken into consideration). The knowledge that can be gained from such nanowires could improve the design of nanowire solar cells and could serve as a bench test to compare the performances of Schottky and p-n diode nanowires.

Also, the ability to synthesize multiple shells around the core allows one to design and synthesize tandem light harvesters (such as P3HT/CdSe/CdS core/shell/shell nanowires).

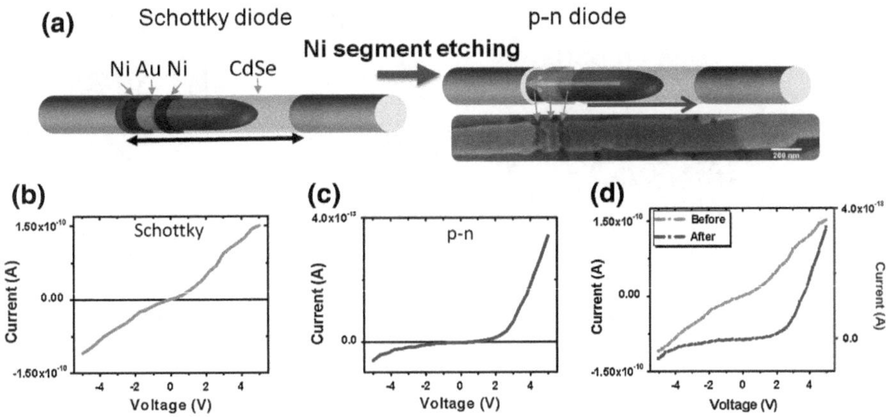

Fig. 6.9 Changing the function of a semiconductor device (preliminary results). **a** Scheme showing the nanowire before and after etching of the Ni segments. Current-voltage characterization of the same nanowire **b** before and **c** after Ni etching. **d** Overlaid curves. For clarity, the axes are different for the two nanowires

6.5 Nanowire Biomimetic Probe Preparation

Insertion of biocompatible probes through the cell membrane is necessary to monitor/modify cell activity. As shown by Almquist et al. [15], insertion of nanowires within the cell membrane is only possible under some specific conditions (Fig. 6.10). They have shown that by replicating the nanometer-scale hydrophilic-hydrophobic-hydrophilic architecture of transmembrane proteins with very fine FIB milling steps, a single artificial probe can be inserted and anchored within the lipid bilayer core [15]. However, the fabrication of such nanowires requires challenging FIB milling steps and only a single nanowire can be prepared at each time. As a simpler and parallel processing alternative, COAL could be used to prepare nanowires composed of an optically active semiconducting core, protected by very thin striped shells of hydrophilic and hydrophobic materials (functionalized nickel and gold segments, for example). Moreover, billions of biomimetic nanowire probes could be prepared in parallel for large-scale cell activity monitoring.

6.6 Photocatalysis

Due to a very short exciton recombination length, conjugated polymers are fairly inefficient materials for photocatalysis. On the other hand, inorganic semiconductors such as CdSe have been found to be very efficient at photogenerating hydrogen under light irradiation in the presence of an electron donor (usually I^-/I_3^- couple).

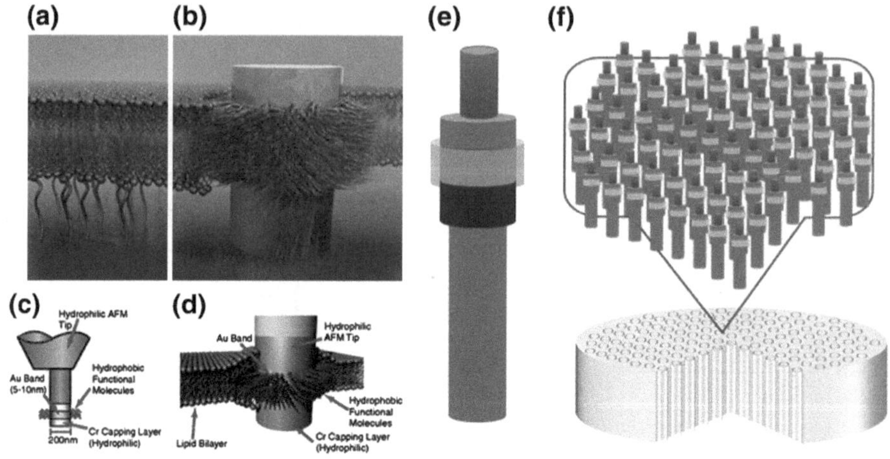

Fig. 6.10 Biomimetic probes (adapted from Almquist et al. [15]). (**a**–**d**) **a** Scheme showing the lipid bilayer attached to the substrate. **b** The hydrophobic part of the probe interacts specifically with the hydrophobic membrane core through selective surface functionalization. **c** Scheme of the nanowire design with nanometers thick functionalized Au surface defined by selective self-assembly of molecules. **d** Scheme of the probe with the hydrophobic core of the lipid bilayer interacting with only the functionalized part, similar to the behavior of membrane proteins. **e** Similar nanowire design, which can be prepared with COAL (selective functionalization step is shown in Fig. 6.7). **f** Scheme showing that such nanowires can be prepared in parallel inside the pores of an AAO membrane, which can be used for large-scale probing of cellular activities

As reported by McFarland et al. [16], multi-segmented PEDOT/CdSe/metal nanowires can be used to photogenerate H_2 (Fig. 6.11a). As a simpler alternative, COAL could be used to generate similar structures in a core-shell fashion, which should greatly improve the efficiency of the structures (Fig. 6.11b). There are ample opportunities provided by COAL to investigate different combinations of materials

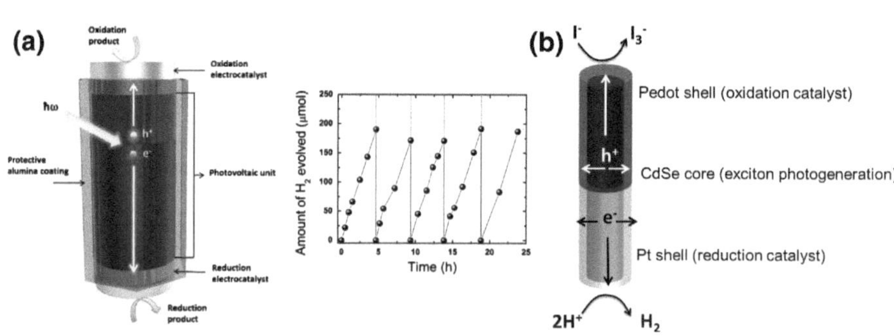

Fig. 6.11 Photocatalysis. **a** Scheme of a multisegmented nanowire (1D stacking of disks) for photocatalysis (*left*) and photogeneration of H_2 by multisegmented nanowires using CdSe as the absorber in a KI electrolyte (pH 2) (adapted from McFarland et al. [16]). **b** Scheme of the core-shell equivalent structure enabled by COAL

and geometries. Insertion of plasmonic rings could also be used to investigate plasmon-enhanced photocatalysis.

In conclusion, I believe that COAL methodology will be useful in wide range of fields including plasmonics, nanomaterial synthesis, biology, photovoltaics, and photocatalysis.

References

1. Qin, L., Park, S., Huang, L., & Mirkin, C. A. (2005). On-wire lithography. *Science, 309*, 113.
2. Wagner, R. S., & Ellis, W. C. (1964). Vapor-liquid-solid mechanism of single crystal growth. *Applied Physics Letters, 4*, 89.
3. Hu, J. T., Odom, T. W., & Lieber, C. M. (1999). Chemistry and physics in one dimension: synthesis and properties of nanowires and nanotubes. *Accounts of Chemical Research, 32*, 435.
4. Hurst, S. J., Payne, E. K., Qin, L., & Mirkin, C. A. (2006). Multisegmented one-dimensional nanorods prepared by hard-template synthetic methods. *Angewandte Chemie, 45*, 2672.
5. Maier, S. A. (2007). *Plasmonics: Fundamentals and applications*. New York: Springer.
6. Kelly, K. L., Coronado, E., Zhao, L. L., & Schatz, G. C. (2003). The optical properties of metal nanoparticles: the influence of size, shape, and dielectric environment. *The Journal of Physical Chemistry B, 107*, 668.
7. Bourret, G. R., Ozel, T., Blaber, M., Shade, C. M., Schatz, G. C., & Mirkin, C. A. (2013). Long-range plasmophore rulers. *Nano Letters, 13*, 2270.
8. Ozel, T., Nizamoglu, S., Sefunc, M. A., Samarskaya, O., Ozel, I. O., Mutlugun, E., et al. (2011). Anisotropic emission from multilayered plasmon resonator nanocomposites of isotropic semiconductor quantum dots. *ACS Nano, 5*, 1328.
9. Ozel, T., Soganci, I. M., Nizamoglu, S., Huyal, I. O., Mutlugun, E., Sapra, S., et al. (2008). Selective enhancement of surface-state emission and simultaneous quenching of interband transition in white-luminophor CdS nanocrystals using localized plasmon coupling. *New Journal of Physics, 10*, 083035.
10. Brittman, S., Gao, H., Garnett, E. C., & Yang, P. (2011). Absorption of light in a single-nanowire silicon solar cell decorated with an octahedral silver nanocrystal. *Nano Letters, 11*, 5189.
11. Hyun, J. K., & Lauhon, L. J. (2011). Spatially resolved plasmonically enhanced photocurrent from Au nanoparticles on a Si nanowire. *Nano Letters, 11*, 2731.
12. Garnett, E. C., Brongersma, M. L., Cui, Y., & McGehee, M. D. (2011). Nanowire solar cells. *Annual Review of Materials Research, 41*, 269.
13. Atwater, H. A., & Polman, A. (2010). Plasmonics for improved photovoltaic devices. *Nature Materials, 9*, 205.
14. Ozel, T., Bourret, G. R., Schmucker, A. L., Brown, K. A., & Mirkin, C. A. (2013). Hybrid semiconductor core-shell nanowires with tunable plasmonic nanoantennas. *Advanced Materials, 25*, 4515.
15. Almquist, B. D., & Melosh, N. A. (2010). Fusion of biomimetic stealth probes into lipid bilayer cores. *Proceedings of the National Academy of Sciences, 107*, 5815.
16. Mubeen, S., Singh, N., Lee, J., Stucky, G. D., Moskovits, M., & McFarland, E. W. (2013). Synthesis of chemicals using solar energy with stable photoelectrochemically active heterostructures. *Nano Letters, 13*, 2110.

Curriculum Vitae

TUNCAY OZEL

12 Oxford St. Office CN308
Cambridge, MA, 02138
tuncayozel@fas.harvard.edu

847-477-5395

CURRENT POSITION

Harvard University, Cambridge, MA

Postdoctoral Fellow in Chemistry and Chemical Biology 2015-present

- **Host:** Prof. Daniel G. Nocera

EDUCATION

Northwestern University, Evanston, IL

Ph.D. in Materials Science and Engineering March 2015

- Thesis Title: Coaxial lithography (Advisor: Prof. Chad A. Mirkin)
- 3M, ECS, SPIE, and University Fellowships (full tuition and stipend)

Bilkent University, Ankara, Turkey

M.S. in Physics September 2009

- Thesis Title: Selective plasmonic control of excitons and their non-radiative energy transfer in colloidal semiconductor quantum dot solids (Advisor: Prof. Hilmi V. Demir)
- SPIE and Board of Trustees Fellowships (full tuition, room, and stipend)

B.S. in Physics May 2007

- Thesis Title: Fabrication and characterization of novel GaN-based optoelectronic devices (Advisor: Prof. Hilmi V. Demir)
- Board of Trustees Fellowship (full tuition, room, and stipend)

RESEARCH EXPERIENCE

Harvard University (Postdoctoral fellow) 2015-present

- **Developed** a novel approach for the preparation of polymer and metal coated silicon nanowires for nanowire assisted drug delivery and nanowire based photonic device applications.

Northwestern University 2010-present

- **Co-invented** a technique bridging templated electrochemical synthesis and lithography, termed coaxial lithography (COAL), for generating coaxial nanowires with sub-10 nanometer resolution in both axial and radial dimensions.
- **Extended** the on-wire lithography (OWL) toolbox with semiconductors to study light-matter interactions. Demonstrated a new type of long-range nanoruler that allows one to probe distances on the 0–100 nm length scale.

T. Ozel, *Coaxial Lithography*, Springer Theses,
DOI 10.1007/978-3-319-45414-6

- **Demonstrated** hybrid semiconductor core-shell nanowires with tunable plasmonic nanoantennas using a versatile method. A photodetector based on a P3HT/CdSe nanowire with a notable on/off ratio was reported.
- **Mentored** and trained two first year PhD students in their projects.
- **Selected** to be leader for the anisotropic materials subgroup (~16 PhD students and post-docs) to lead weekly subgroup meetings, prepare grant proposals and reports, develop collaborations, and mentor new students joining the subgroup.

Bilkent University 2005-2010

- **Co-invented** an optoelectronic device for secure communications in UV that can serve as a quantum modulator, detector, and light source on a single chip.
- **Co-invented** fluorine based white light emitting polymers for use in hybrid light emitting diodes.
- **Co-invented** a photocatalytic nanocomposite material for use in self-cleaning surfaces.
- **Demonstrated** the ability to selectively enable/disable plasmon-exciton formation distinctly at the donor site or at the acceptor site of a quantum dot pair interacting each other *via* Förster-type resonance energy transfer.
- **Reported** anisotropic emission and record quantum efficiency for CdTe quantum dot solids by creating highly strong plasmon-exciton interactions in a nanocomposite localized surface plasmon resonator.

SCHOLARSHIPS & AWARDS

Turkish American Scientists and Scholars Association (TASSA) Young Scholar Award	2016
Electrochemical Society (ECS) Edward G. Weston Fellowship	2014
International Institute for Nanotechnology (IIN) Outstanding Researcher Award	2014
3M Science and Technology Fellowship	2012-2014
SPIE Optics and Photonics Education Scholarship	2013
Materials Research Society (MRS) Fall Meeting Graduate Student Award	2013
American Physical Society (APS) Travel Grant for Opportunities in Energy Research Workshop	2013
Northwestern University Department/Research Assistant Fellowship	2010-present
Education USA Opportunity Grant	2009
SPIE Optical Science and Engineering Scholarship	2008
Bilkent University Board of Trustees Fellowship for BS and MS studies	2003-2009

PATENTS

1. C. A. Mirkin, **T. Ozel**, G. Bourret, Coaxial lithography, application no: US62/000861 (2014).
2. G. Celiker, H. Demir, F. Ozkadi, S. Tek, E. Mutlugun, I. Soganci, **T. Ozel**, I. Huyal, A photocatalytic nanocomposite material, PCT-WO2009133010, CN102164860, JP2011519720, EC-EP2294014 (2008).
3. H. Demir, D. Tuncel, I. Huyal, **T. Ozel**, White emitting polymer emitters for white LEDs, TR200809254 (2008).
4. H. Demir, **T. Ozel**, E. Sari, Quantum modulator, detector and light source on a single chip for communications in UV, TR200708850 (2007).

ACADEMIC PUBLICATIONS (h-index: 16, *equal contribution)

1. **T. Ozel***, M. Ashley* *et al.* Solution dispersible metal nanorings: Independent control of architectural parameters and materials generality, ***Nano Letters*** 15 (8), 5273 (2015).
2. **T. Ozel***, G. Bourret* *et al.* Coaxial lithography, ***Nature Nanotechnology*** 10, 319 (2015).
3. K. Osberg, N. Harris, **T. Ozel** *et al.* Systematic study of antibonding modes in gold nanorod dimers and trimers, ***Nano Letters*** 14 (12), 6949 (2014).
4. S. V. Zhukovsky, **T. Ozel** *et al.* Hyperbolic metamaterials based on quantum-dot plasmon-resonator nanocomposites, ***Optics Express*** 22 (15), 18290 (2014).
5. **T. Ozel***, G. Bourret* *et al.* Hybrid semiconductor core-shell nanowires with tunable plasmonic nanoantennas, ***Advanced Materials*** 25, 4515 (2013).
6. **T. Ozel***, P. L. Hernandez-Martinez* *et al.* Observation of selective plasmon-exciton coupling in nonradiative energy transfer: Selective plexcitons, ***Nano Letters*** 13 (7), 3065 (2013).

7. G. Bourret*, **T. Ozel*** *et al.* Long-range plasmophore rulers, ***Nano Letters*** 13 (5), 2270 (2013).
8. M. Rycenga, M. Langille, M. Personick, **T. Ozel** *et al.* Chemically isolating hot spots on concave nanocubes, ***Nano Letters*** 12 (12), 6218 (2012).
9. **T. Ozel** *et al.* Anisotropic emission from multilayered plasmon resonator nanocomposites of isotropic semiconductor quantum dots, ***ACS Nano*** 5 (2), 1328 (2011).
10. U. O. S. Seker, **T. Ozel** *et al.* Peptide mediated constructs of quantum dot nanocomposites for enzymatic control of non-radiative energy transfer, ***Nano Letters*** 11 (4), 1530 (2011).
11. I. O. Ozel, **T. Ozel**, *et al.* Non-radiative resonance energy transfer in bi-polymer nanoparticles of fluorescent conjugated polymers, ***Optics Express*** 18, 2, 670 (2010).
12. E. Mutlugun, O. Samarska, **T. Ozel** *et al.* Highly efficient nonradiative energy transfer mediated light harvesting using aqueous CdTe quantum dot antennas, ***Optics Express*** 18, 10, 10720 (2010).
13. N. Cicek, S. Nizamoglu, **T. Ozel** *et al.* Structural tuning of color chromaticity through nonradiative energy transfer by interspacing nanocrystal monolayers, ***Applied Physics Letters*** 94, 061105 (2009).
14. **T. Ozel** *et al.* Selective enhancement of surface-state emission and quenching of interband transition in white-luminophor CdS nanocrystals using localized plasmon coupling, ***New Journal of Physics*** 10, 083035 (2008).
15. I. O. Huyal, **T. Ozel** *et al.* Quantum efficiency enhancement in film by making nanoparticles of polyfluorene, ***Optics Express*** 16, 17, 13391 (2008).
16. E. Sari, **T. Ozel** *et al.* Comparative study of electroabsorption in InGaN/GaN quantum zigzag heterostructures with polarization-induced e-fields, ***Applied Physics Letters*** 92, 201105 (2008).
17. I. O. Huyal, **T. Ozel** *et al.* White emitting polyfluorene functionalized with azide hybridized on near-UV light emitting diode for high color rendering index, ***Optics Express*** 16, 2, 1115 (2008).
18. I. O. Huyal, U. Koldemir, **T. Ozel** *et al.* On the origin of high quality white light emission from organic/inorganic LED using azide functionalized polyfluorene, ***Journal of Materials Chemistry*** 18, 3568 (2008) (**inside cover**).
19. S. Nizamoglu, E. Mutlugun, **T. Ozel** *et al.* Dual-color emitting CdSe-ZnS heteronanocrystals hybridized on InGaN/GaN LEDs for high-quality white light generation, ***Applied Physics Letters*** 92, 113110 (2008).
20. H. V. Demir, S. Nizamoglu, E. Mutlugun, **T. Ozel** *et al.* Tuning shades of white light with multi-color quantum-dot-quantum-well emitters, ***Nanotechnology*** 19, 335203 (2008).
21. **T. Ozel** *et al.* Violet to deep-ultraviolet InGaN/GaN and GaN/AlGaN quantum structures for UV electroabsorption modulators, ***Journal of Applied Physics*** 102(11), 113101 (2007).
22. S. Nizamoglu, **T. Ozel** *et al.* White light generation using CdSe/ZnS core-shell nanocrystal hybridized with InGaN/GaN light emitting diodes, ***Nanotechnology*** 18, 065709 (2007) (**cover**).
23. H. V. Demir, S. Nizamoglu, **T. Ozel** *et al.* White light generation tuned by dual hybridization of nanocrystals and conjugated polymers, ***New Journal of Physics*** 9, 362 (2007).
24. E. Sari, S. Nizamoglu, **T. Ozel** *et al.* Blue quantum electroabsorption modulators based on reversed quantum confined Stark effect with blue shift, ***Applied Physics Letters*** 90, 011101 (2007).

SELECTED CONFERENCE TALKS

1. **T. Ozel** *et al.* Synthesis of solution-dispersible core-shell semiconductor nanowires and metal nanorings with deliberately controllable compositions and architectural parameters, 2015 MRS Fall Meeting, Boston, MA, USA (December 2015).
2. **T. Ozel** *et al.* Coaxial lithography, 2014 MRS Fall Meeting, Boston, MA, USA (December 2014).
3. **T. Ozel** *et al.* Plasmon coupled hybrid semiconductor nanowires: synthesis and applications, 2014 MRS Spring Meeting, San Francisco, CA, USA (April 2014).
4. **T. Ozel** *et al.* Hybrid semiconductor core-shell nanowires with tunable plasmonic nanoantennas, 2013 MRS Fall Meeting, Boston, MA, USA (December 2013).
5. **T. Ozel** *et al.* Observation of anisotropic emission from semiconductor quantum dots in nanocomposites of metal nanoparticles, Proceedings of 23rd IEEE Annual Photonics Society Meeting, Denver, CO, USA (November 2010).

6. **T. Ozel** *et al.* Plasmon enhanced colloidal nanocrystal composites incorporating Au nanoparticles in a repeating layered architecture, Proceedings of IEEE Photonics Society Annual Meeting, Antalya, Turkey (October 2009).
7. **T. Ozel** *et al.* Surface-state emission enhancement in white-luminophor CdS nanocrystals using localized plasmon coupling, Proceedings of IEEE Lasers and Electro-Optics Society Meeting, Newport Beach, CA (November 2008).
8. **T. Ozel** *et al.* Near-UV InGaN/GaN-based quantum optoelectronic devices, Proceedings of SPIE Europe Microtechnologies for the New Millennium, Maspalomas, Gran Canaria, Spain (May 2007).

RESEARCH SKILLS

Fabrication

- Cleanroom Class-100/1000 experience: Photolithography, electron beam lithography, thermal and electron beam evaporation, RF sputter coating, reactive ion etching, wet etching, plasma enhanced chemical vapor deposition, rapid thermal annealing.
- Nanoparticle synthesis experience: Electrochemical synthesis of metal and semiconductor nanowires (including core/multi-shell nanowires), colloidal quantum dot nanocrystal synthesis (CdSe, CdTe, CdSe/ZnS core/shell), metal nanoparticle synthesis (Au, Ag, Au/Silica core/shell, Ag/Silica core/shell).

Imaging and characterization

- Transmission electron microscopy, scanning electron microscopy, energy dispersive X-ray spectroscopy, atomic force microscopy, time-resolved (picosecond) and steady-state fluorescence spectroscopy, UV-Vis spectrophotometry, confocal microscopy, semiconductor electrical parameter analysis.

Software

- Lumerical for finite-difference time-domain (FDTD) simulations, FluoFit for fluorescence decay lifetime analysis, Matlab, LabVIEW for instrument control, DesignCAD for electron beam lithography pattern design, AutoCAD for photomask design.

TEACHING EXPERIENCE

Teaching Assistant of Graduate Level Courses (office hours and grading)

- Solar Energy Conversion — Spring 2013
- Advanced Physics of Materials — Fall 2011
- Experimental Methods in Applied Physics — Spring 2009, Spring 2010
- Nanoscience and Nanotechnology — Fall 2009

Laboratory Assistant of Undergraduate Level Courses (demonstrations and grading)

- Experimental Physical Chemistry (including group project mentoring) — Spring 2016
- Introduction to Mechanics — Fall 2007, Fall 2008, Fall 2009
- Introduction to Electricity & Magnetism — Spring 2007, Spring 2008, Spring 2010

LEADERSHIP EXPERIENCE

Guest Lead Editor of Journal of Nanomaterials, Hindawi Publishing Corporation — 2015-2016

Leader of anisotropic materials subgroup, Mirkin Group, Northwestern University — 2013-2015

Vice President of Turkish intercultural club, Northwestern University — 2012-2014

Graduate Chair of Turkish student association, Northwestern University — 2013-2014